The Science of Story

The Science of Story

The Brain Behind Creative Nonfiction

Sean Prentiss and Nicole Walker

BLOOMSBURY ACADEMIC
LONDON • NEW YORK • OXFORD • NEW DELHI • SYDNEY

BLOOMSBURY ACADEMIC
Bloomsbury Publishing Plc
50 Bedford Square, London, WC1B 3DP, UK
1385 Broadway, New York, NY 10018, USA

BLOOMSBURY, BLOOMSBURY ACADEMIC and the Diana logo are trademarks of Bloomsbury Publishing Plc

First published in Great Britain 2020

Cover design: Eleanor Rose
Cover image © Getty Images

A catalogue record for this book is available from the British Library.

Library of Congress Cataloging-in-Publication Data
Names: Prentiss, Sean, editor. | Walker, Nicole, editor.
Title: The science of story: the brain behind creative nonfiction / edited by Sean Prentiss, Nicole Walker.
Description: London; New York: Bloomsbury Academic / Bloomsbury Publishing, Plc: 2020. | Includes bibliographical references and index.
Identifiers: LCCN 2019019128 (print) | LCCN 2019980168 (ebook) | ISBN 9781350084247 (paperback) | ISBN 9781350083905 (epub) | ISBN 9781350083899 (pdf)
Subjects: LCSH: Creative nonfiction–Authorship. | Essay–Authorship. | Literature and science. | Storytelling–Psychological aspects. | Neurosciences and the humanities.
Classification: LCC PN145 S35 2020 (print) | LCC PN145 (ebook) | DDC 808.02–dc23
LC record available at https://lccn.loc.gov/2019019128
LC ebook record available at https://lccn.loc.gov/2019980168

ISBN: HB: 978-1-3500-8388-2
PB: 978-1-3500-8424-7
ePDF: 978-1-3500-8389-9
eBook: 978-1-3500-8390-5

Typeset by Deanta Global Publishing Services, Chennai, India

Contents

1

Introduction

Sean Prentiss and Nicole Walker

As professors of creative nonfiction, we—Sean and Nicole—teach students to write creative nonfiction. We ask students to consider scene and exposition, to interrogate the complexities of memories, and to consider time on the page. But, creative nonfiction is also one of the most epistemologically pressing art forms. How can we know what we know? What constitutes truth—fact, experience, or consensus? By the mere act of trying to represent ourselves on the page, have we already begun to dissemble? Has that memory we have turned into "scene" to show our story replaced an earlier, truer memory? What form best reflects a particular experience? Does lyric trump narrative or flashback? Can a hermit crab essay best describe how multiple perspectives reflect a common story?

As parents, we are used to hearing our children say, "But why?" But this question is not only a question that children ask their parents. It is also a question that any good student should ask of their teachers. And our students often ask us that exact question. But what if the teacher doesn't have the answer? *The Science of Story* was born out of our desire to understand not only *what* we should do on the page (and what we should teach others to do) but also *why* we should do those things.

The Science of Story strives to understand that question of "But why" by examining what cognitive science (the study of thoughts), neuroscience (the study of the brain), and other sciences might teach us about creative nonfiction and about the human experience. Our sixteen writers delve into recent science to learn how science supports, works against, or adds texture to techniques creative nonfiction writers use. These writers tie this science directly to creative nonfiction and offer ideas on how we continue to evolve and grow as effective creative nonfiction writers.

The Science of Story investigates, essay by essay, foundational ideas of creative nonfiction and searches for answers to questions that surround creative nonfiction. These essays examine how creative nonfiction works (and how the human mind works) and offer suggestions on how we can understand and use this science to improve, complicate, or expand our writing lives, our craft, and our teaching of writing.

Each essay uses research to create craft essays that live within the Venn diagram that is the human experience, creative nonfiction, and science. Some essays blend the personal with research, while others highlight how cognitive and neuroscience illuminate how and why we write the way we write. Some others use science as a metaphor to examine our writing and our writing lives. All of these pieces are written by creative nonfiction writers who, either through schooling or personal interest, are deeply engaged in exploring how cognitive and neuroscience work in conjunction with creative nonfiction.

This book, in many ways, serves as a *how-to* (by offering suggestions for writers to continue to improve their craft), a *why* (explaining why we might consider doing things, based not on observation but on cognitive and neuroscience research), and a *what-if* (exploring what other ideas might be possible). All of these craft essays are based upon cognitive and neuroscience but explained through the lens of creative nonfiction.

But this book refuses to be prescriptive in telling writers how science commands us to write. Rather, this anthology offers new techniques (or proves old techniques) while also allowing readers to envision new tools to add to their writing lives. Still, what we are most excited about this collection is that these essays ruminate on what science can teach us about what it means to be writers trying to understand and portray the creative nonfiction world of our lives.

2

Bengal Tiger Moments: Perception of Time in the Brain and on the Page

Sean Prentiss

Creative nonfiction is not just a recording of the events of our lives—car accidents, swooning love, the slow building of a remote cabin—but it is also how we remember, interpret, and re-create these events. We have the great big world out there—of people and places and things—and, completely removed, we have our brains cocooned by skulls. Our brains, literally, are shrouded in darkness. The only way our brains reach past our skulls to experience the outer world is through electrical signals speeding down bundled nerves from our eyes, noses, ears, skin, and tongue to our brains. Without these electrical signals, our brain's understanding of the world would be less than black.

Through the brain's processing of the senses, our brains get to live outside of themselves. They get to experience and interpret the larger world. And as our brains experience that outer world, they are forced to perceive time and to use this perception of time to link moments of life into created memories of events. And this is no small thing. Time perception is so important that our brains are born understanding the passage of time. According to psychologist Sylvie Droit-Volet, newborns possess a "primitive" understanding of time and infants as young as one month old react based on the movement of time.[1]

And though humans are born understanding and reacting to time, scientists are unsure where the brain perceives time. Most scientists, including Marc Wittman, a scientist at the Institute for Frontier Areas of Psychology and Mental Health, in Freiburg, Germany, believe that "There is no one area of the brain, or system in the brain, that is solely dedicated to recording the passage of time." Instead, time related processes are distributed among the cerebral cortex, cerebellum, and basal ganglia, among others, and the brain has a variety of "clocks" that handle different types of brain time including millisecond clocks used for rapid reaction, decade clocks that deal with the physical body, and clocks that deal with memory or predicting the future.[2] Without these clocks, humans could not survive. We couldn't breathe on a scheduled pattern, eat when we needed to, or sleep enough. We'd misplace or run out of or fall through time and die.

But even though we need to perceive time to survive, and though we could argue that time moves simply enough: one second chronologically lasts as long for me as it does for you—the reader of this essay—still, perception of time can vary wildly. Time can appear to screech to a halt when we're bored or worried. And, as clichéd as it sounds, time appears to fly when we are having fun. So although time is universal, perception of time is a construct of the brain. And our perception of time's movement is faulty much of the time. Rather than feeling as if a one-minute experience lasted one minute we often feel as if one minute lasted up to a third shorter or longer.

This faulty perception affects not just our lives but also our writing lives because how we perceive the world around us affects the details we remember and later write in our creative nonfiction. Our nonfiction is most affected by those moments that were so important that they slant our memories. And as we writers realize that these moments are powerful enough to affect our brain's understanding of time, we can use those memoires on the page by lingering in those rich details to make our essays and memoirs erupt with life.

These richest moments arise from three neurological issues. The first issue is fight or flight related experiences. When we live through high-octane moments—like a first kiss in a kitchen in Wayne, Pennsylvania, or the slow spin of a truck on an icy Vermont road, veering toward a tree—time seems to stretch forever before the two lips actually touch or the truck kisses that old birch.

Why does time slow during fight or flight moments?

It doesn't.

It only *feels* as if time grinds to a halt. David Eagleman, neuroscientist and author of *The Brain: The Story of You* and *Incognito: The Secret Lives of the Brain*, performed an experiment to see if time slows during fight or flight moments. Eagleman had people do unexpected freefalls from multiple stories into a safety net. During this terrifying fall, Eagleman asked his subjects to focus on reading flashing numbers on a perceptual chronometer that showed numbers barely faster than humans can perceive. If time actually slowed during this freefall, the faller would be able to read the numbers on the chronometer; if time didn't slow, the faller wouldn't be able to read the numbers. During all the freefalls, not one person could read the chronometer.

Afterwards, Eagleman asked his subjects to "re-create your freefall in your mind. Press the stopwatch when you are released, then press it again when you feel yourself hit the net." Eagleman found that fallers' "estimates of their own fall were a third greater, on average, than their recreations of the fall of others."[3] People thought the amount of time they were falling was a third longer than it really was.

Scientists suspect time appears to slow because it is evolutionarily beneficial. In moments of danger—let's say a tiger enters the room where you are reading this essay, a Bengal tiger with a light orange coat with white and black stripes, ten feet long from his four-inch-long fangs to his flicking tail, nearly four feet tall at the muscled shoulders, and close to 550 pounds. What do you do? Do you quietly notice the tiger and then keep reading? Do you call to your partner in the other room? *Excuse me, but do you see the tiger by the door? I believe it's a Bengal tiger.*

No, you run.

You fight for the door.

You race past whoever is in the house with you.

You cower in the corner, too frightened to even move.

Or you use your copy of *The Science of Story* as a shield and a nearby pen as a knife, ready to fight to the death.

When we experience fight or flight moments, according to Eagleman, "the amygdala kicks into high gear, commandeering the resources of the rest of the brain and forcing everything to attend to the situation at hand."[4] Brains kick into a higher gear, Wittman adds, "Because of the threatening situation, I am totally aroused, and my internal physiological processes speed up. And so relative to that, the outside world slows down."[5] Therefore, time appears to slow as the brain maximizes the amount of information it accesses through our senses, so we experience a heightened perception so our brains can more intelligently make decisions.

When the amygdala is engaged, not only do our brains work at a higher level, but our memories are also laid down in a way which provides, again, according to Eagleman, "the later flashbulb memories of post-traumatic stress disorder [and PTSD is always born out of fight or flight related moments]. So in a dire situation, your brain may lay down memories in a way that makes them 'stick' better. Upon replay, the higher density of data would make the event appear to last longer."[6]

Fight or flight related writing can be seen in Edward Abbey's essay, "Havasu" from *Desert Solitaire*. Abbey spent a day in 1949 wandering down Havasu Canyon, slowly lowering himself down the waterfalls until he dropped off one fall and, Abbey writes:

> I hit rock bottom hard, but without any physical injury. I swam the stinking pond dog-paddle, pushing the heavy scum away from my face, and crawled out on the far side to see what my fate was going to be.
>
> Fatal. Death by starvation, slow and tedious. For I was looking straight down an overhanging cliff to a rubble pile of broken rocks eighty feet below.[7]

Abbey finds himself caught on the edge of a waterfall, with a fall to death below and no way to climb out of the pool above. For the next page, Abbey writes about his ideas on how to escape—screaming for help, tearing his clothes into strips, and building a signal fire. Once he realizes those won't work, Abbey tries scaling the vertical rock wall:

> Here I was able to climb upward, a few centimeters at a time, by bracing myself against the opposite sides and finding sufficient niches for fingers and toes. [. . .] Somehow, with a skill and tenacity I could never have found in myself under ordinary circumstances, I managed to creep straight up that gloomy cliff and over the brink of the drop-off and into the flower of safety. [. . .] I discovered myself bawling again for the third time in three hours, the hot delicious tears of victory. And up above the clouds replied—thunder.[8]

Abbey published "Havasu" in *Desert Solitaire* in 1968, nearly twenty years after the near death experience occurred. Yet, Abbey's recollection is pristine and seared into his memory. And Abbey elongates this short moment stuck on a cliff in Havasu Canyon for nearly five pages, highlighting the power of fight or flight moments to stretch time.

Alongside fight or flight related issues, novel experiences also affect how we perceive time. Our perception of time speeds up as we age because for a young child, every experience is either new or was recently new. During our first days of attending school, we ask: How do we get on the school bus and where do we sit? When do we go to the bathroom? Who will be our friend? As with fight or flight related issues, these novel experiences get recorded more richly because every experience is important to our survival; children devote great amounts of brain power to learning how to live in the world.

To a middle schooler, school has become old hat. To a high school senior, school might be the least interesting place imaginable. Once you're my age, 48, you've spent 2,500 days at school as a student and another 2,000 as a professor. Now almost nothing new happens. I commute the same roads to work (County Road to Main Street to Route 12). I teach the same Norwich University students. Then I go home. It's not that I don't love my job. I do. It's just that since so few novel experiences happen each day (each week, month), I don't use much brain power to figure out how to survive at school. The world has become mundane enough that the brain "forgets" most of the new information it accumulates each day because this information is a repetition of what was learned before. That's why I forget what I wore to teach last week but remember what I wore to my first day of elementary school (blue slacks and a blue striped shirt and a bowl haircut).

Before we examine our third issue that skews time perception, let us, for a moment, take a break from thinking about these time related ideas and travel to the Hope Zoo in Kingston, Jamaica, in 1996. It is, as always, a glorious day at the zoo. Families picnic in the gardens. A Caribbean sun shimmering down like diamonds. And we wander around looking at the animals—the ring-tail coati, the pacca, the serval—before we stop in front of the big cats cages. These cages are just that, cages that we can walk right up to. So that is what we do. We walk to within a foot of the cage, which is by far the closest we've ever gotten to a big cat.

The big cats—tiger and lions in their individual cages—are lying down, their backs against cage bars, heads away from us. They flick tails intermittently. We are so close that if we reach out our left hand, we might be able to touch the tiger's tail. So we do that. We stretch our hand out, slowly inching it closer. We are behind the tiger's back so it cannot see us, we assume. When our hand is mere inches from the tail, the tiger—without looking at us—lets out a low, rolling guttural growl that might mean *I will rip these bars down*. We trip over our feet, stumbling backwards. And this moment forever remains etched into our minds even though earlier today and later today some anonymous zoo keeper will bring food to this cage, walk right up (or maybe even *into* the cage) to feed this tiger the same way you or I might feed our dog. One moment is never forgotten because it is new; the other moment is forgotten because it is routine.

An example of a novel moment in writing is in Charles Bowden's devastating essay on humans and violence, "Torch Song." Bowden, at the beginning of "Torch Song," introduces us to the first time he covers a violent crime as a news reporter. This novel experience is rich and sick and brutal and filled with the exact details we expect from a novel experience:

I can't tell much from her silhouette. She's sitting off to one side, her shoulders hunched, and toward the front is the box with the teddy bears. Or at least I think they're teddy bears. Almost twenty years have passed, and I've avoided thinking about it. There are some things that float pretty free of time, chronology, the book of history, and the lies of the experts. In the early Eighties I went to a [a motel where a child had been murdered] as part of my entry into a world, a kind of border crossing.

It started as the golden light of afternoon poured through the high, slit windows of the newsroom. I had no background in the business and I'd lied to get the job. I was the fluff writer, the guy brought on to spin something out of nothing for the soft features [. . .] One afternoon the newsroom was empty, and the city desk looked out and beckoned me. I was told to go to a motel and see if I could find anything to say.

[. . .] The motel I was sent to was a hot-sheet joint, with rooms by the hour or day, and featured water beds (WA WA BEDS, in the language of the sign), in-room pornographic movies, and a flock of men and women jousting through nooners.

I walked around aimlessly [inside the motel room] and popped open the door of the old refrigerator—shelves empty—and then the little door to the freezer, where two bottles of Budweiser, frozen solid, nestled as if someone with a powerful thirst had placed them to chill in a hurry and then been distracted. I heard the woman's voice in my ear explaining how the mother had gone to work—she danced at a strip joint, one of the new gentlemen's clubs that featured college-looking girls instead of aging women with bad habits—and so was gone when it happened. [. . .] I looked at a big splotch on the cinderblock wall, and she said, "I haven't had time to clean that off yet."

> That's where the head had hit, the skull of the toddler just shy of two years, as the man most likely held him by the legs and swung him like a baseball bat. He probably killed the kid out of boredom or frustration with the demands of a small child, or because he'd been bopped around himself as a child, or God knows why.[9]

Later in "Torch Song," Bowden comes to live, for years and years, on the edge of this world of violence and deviancy and sex. He lives a world where reporting on violence becomes an everyday occurrence. These brutal moments lose their power to Bowden. And as they lose their power, Bowden falls deeper into a world of deviant sex. Rather than these deviant sexual moments gaining power, they lose power because he has experienced so much. Bowden writes:

> A woman is at the door and she says she has cuffs.
>
> A woman is at the door late at night and we make love, and as she leaves she says she can't see me again because she is getting married in the morning.
>
> Two women are at the door . . .[10]

By the end of "Torch Song," rape, murder, child molestation, and risky sex become the norm. What was once shocking and novel and powerful becomes nearly, somehow, mundane.

Outside of fight or flight and novelty experiences, we have a third issue that affects perception of time: altered states. This can include drug and alcohol use, dementia, psychological issues (like bipolar disorder), spiritual issues, and medical issues (we call it *chemo brain* for a reason).

Altered states experiences affect perception of time because stimulants like cocaine and meth or depression or anxiety make us overestimate how long an event takes while depressants like alcohol and marijuana or maniac moments make us think an event went by more quickly. Psychoactive drugs like LSD and mescaline may make time speed up, slow down, or become out of sequence.

Imagine a person sitting in their living room. Then they take a psychoactive drug of some sort and sit on their old and weary couch. They look at the painting on their wall. Maybe it is Andrew Wyeth's *Dorothy's World*. Our drug user stares at the image of Dorothy, propped up on one hip in a sea of tan grass, a farmhouse off in the distance. As the psychoactive drugs take effect, our person hears something rustling Dorothy's grass, something that is moving closer and closer to Dorothy, who seems stuck to the ground and unable to run away. The grass parts enough so our drug-affected individual sees the orange, white, and black markings of a Bengal tiger pushing through the dried grass, stalking Dorothy. The tiger draws its face back into a roar that lasts so long that our drug user can do nothing but sit paralyzed on their couch. After what seems like minutes or hours, the tiger lunges toward Dorothy in a blur. Right before landing upon her—our drug user screaming—the tiger vanishes and Dorothy is again alone in her field of dried grass and our drug user is left on the couch, dazed and scared.

An example of drug related experiences can be seen in Jessica Hendry Nelson's memoir, *If Only You People Could Follow Directions*, which deals with Nelson's family's struggle with addiction and mental illness, Nelson writes about an evening and night where she and her friend Jordan drink wine and snort Ritalin:

> We swig back glasses of wine and light cigarettes. "Let's take a walk," I say.
>
> "I can't move," says Jordan. "I mean, I just want to be right here, with you."
>
> We sit quietly, smoking, and I feel my thoughts begin to trip over one another, my heart racing. I watch the old flip clock clicking

> through those arbitrary numbers, and it seems so loud, that clicking [. . .]
>
> Jordan reaches over and covers the clock with a blanket. It goes silent, which we soon realize is not the same as stopping time.[11]

Soon, it is 2:30 am and Jordan is crying as their friend Angel visits. Then time passes in one big leap.

> Here I am, and that sun, which isn't here yet, could show up at any moment unannounced, as it tends to do. The world will start up its endless, painstaking rotations, and all the people will get out of bed. [. . .]
>
> "Oh my God," I say. "Do you know what that means? It means it's late, really late, and soon the sun will come up. I can't bear to see the sun. God, I hate that fucking sun."[12]

In this section, we get a scene where time moves in jumps and starts. It lingers in moments that stretch out and then it contracts as time leaps forward from evening to the middle of the night to the edge of almost dawn. During moments of drug usage, time is not linear but is instead a jumping, fidgety thing, an elusive thing that yawns and accordions and disappears.

Now that we know the major experiences that make brains misperceive time, we can consider how to write these moments in ways that cause readers to also misperceive time. Two ways to help the reader feel the warping of time are by using speed of scene and incident frequency.

Speed of scene is the amount of time it takes a reader to read a scene versus the amount of time the actual event took. Five categories exist for speed. From fastest to slowest they are Gaps, Summary, Scene, Dilation, and Internal Scene.

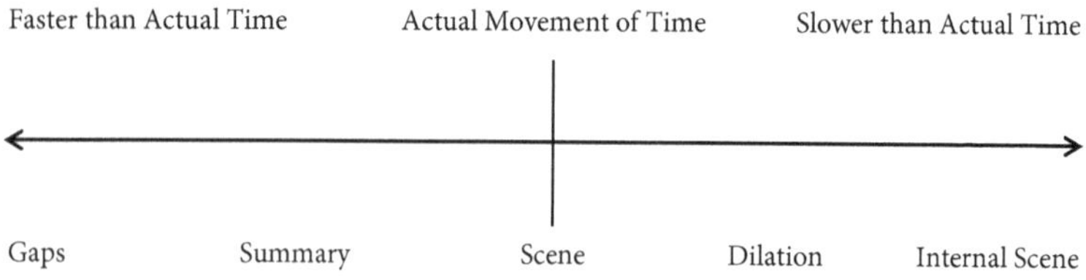

Gaps are the fastest movement of time. Gaps, as their name implies, leave out many details of an event. Gaps skip over age related moments where we forget things because they are unimportant. Drug-affected moments often have gaps.

Erik Reece, in *Lost Mountain*, offers an example of a gap when in "Acts of God" he writes,

> In 1912, the railroad finally reached McRoberts, a small hamlet that sits near the headwaters of the Kentucky River's north fork. The Northern coal barons had been waiting for decades to get at the minerals in the Cumberland Highlands, and once the tracks were laid, they quickly threw up coal camps along the narrow valley floors. Elkhorn Coal built nearly a hundred four-over-four houses about ten feet from either side of the only road that leads to McRoberts. Two families lived in each house. There was no indoor plumbing.
>
> Men who had tended their marginal farms traded plows for picks and went to work in the new mines. They made enough to pay rent to the coal company and buy canned food at the company store. The latter would become a symbol of the miner's loss of independence. "Though he might revert on occasion to the ancestral agriculture," wrote Harry Caudill, "he would never again free himself from dependence upon his new overlords." It was in a very real sense that the miner narrating the song "Sixteen Tons" felt that he owed his soul to the company store.
>
> Today, many of the company houses are still standing.[13]

In one line of text, Reece gaps us from 1912 to "Today." Why does Reece skip nearly one hundred years? Because those years had nothing to do with Lost Mountain or with mountaintop removal. Hence, the use of a gap that spans multiple years.

Summary, much like gaps, deals with forgettable nonnovel moments. When using summary, we distill an event down into just important details. Summary can compress months of dating into just a few short lines of text.

An example of summary can be seen in Jane Alison's *The Sisters Antipodes*. In the scene below, Jane is invited to Washington D.C. to have dinner with her former stepfather, Paul. Paul asks Jane how her mother, Paul's former wife, is doing. Here Alison shifts us from scene to summary.

> Another pause and jingle of ice, another shift of subject.
>
> "Your mother still seeing—what's his mane—Phil? Bill? [asks Paul of Jane]."
>
> She probably was, Phil or Bill or Bob or Ken or some John or another. Men pulled up in white Cadillacs on Friday and Saturday and took her to dinner at Picadilly, she would dance out all Givenchy, leggy blue dress, bright gap-toothed smile, and what was so hard to work out, as I sat on Paul's slick black chair and dug my feet into that hateful white carpet and looked at all those bright brown eyes [of Paul and his children] and laughing straight American teeth, was the right way to answer, how to know whether my mother would have a positive valence for attracting these men or a negative valence for needing to.[14]

In this excerpt, Alison begins in scene but then moves to summary to skim over details that could be turned into pages of scene in a memoir about the dating history of Jane's mother. Instead, Alison summarizes the men who are entering her mother's life because what matters is not that we come to know each suitor but that we come to know the long line of men as a unit, as a group of unimportant men who enter and leave Jane's life.

Scene matches the actual time of the event. This is closest to the real time of an event. Here the time it takes to read a narration closely matches the actual time of the event. Scenes often come from all the times when our perception is affected—fight or flight, novel, and altered state experiences.

Joy Castro, in *Island of Bones*, offers one example of a move from summery to scene. Our narrator, Joy, tries to convince her mother not to remarry.

> I cry in her room as she dresses, begging her not to do it, but I have no evidence aside from the weird way he looks at us. She's patient for a while, going over the money he makes, the good reputation he has in his congregation—but finally she turns on me.
>
> "I am just about fed up, you hear? Do you understand me? I've just about had it with your bellyaching." She swings the hairbrush in my face. "Why do you always want to ruin everything? Why? One good thing comes along, something that will actually make me happy for once, and you have to start your whining. As usual."
>
> "He's not a good man." I'm still crying. She laughs angrily, throws the brush down on the bureau.[15]

Castro begins this except closer to summary ("I cry" and "begging her not to" and "She's patient for a while") before moving into the moment that is seared into Castro's brain due to the power of the event—the anger bursting from her mother, the threat of the hairbrush in Joy's face, the painful dialog.

Dilation occurs when the reading of a moment stretches longer than the actual event. Dilation is slower—because of the level of detail added—than real life and often mirrors fight or flight, novel, and altered state experiences.

Charles Finn, in his collection of essays, *Wild Delicate Seconds*, dilates a moment when he gazes upon a cougar. In real life, the cougar takes one single step while Finn is watching, a moment that lasts less than a second. From that one second, Finn writes:

> The cougar stops and stands in profile. Large chamomile eyes swivel and loop up into mine. Its tail is drooped in an elongated S, shoulders rounded above the sweep of its back. Powder-puff ears turned toward me and its face is centered around the pink heart of a nose. I can make out the thick lines of its whiskers, also the indent of lips pulled back in their perpetual smile. The cougar lifts its head and dignity oils its fur. It raises a boxing glove paw as if in greeting and I know my one wish above all others is to spend time with one of these cats, to hold and pet one, to hear one of them purr.
>
> The cougar sets its paw down ahead in the snow. It lifts the other.[16]

Finn dilates time so much that we get nine sentences describing a cougar taking a single step. Finn dilates because this fight or flight moment was so powerful that his five scenes were able to share more information with the brain and the brain recorded the memory more richly.

Internal scene is when a writer stops action and moves from the real world into the brain for thought. Because the writer is no longer experiencing the outer world, there is no perception of time occurring, instead only introspection, speculation, and reflection.

An example of a move from scene to pause to scene is when Joe Wilkins, in *The Mountain and the Fathers*, visits Ed Dempsey to get a scent to help him trap coyotes.

> Ed Dempsey is big as a bear, but slumped and pudgy and bald. He has turned his big body away from me, is studying the rows of shelves along the far wall. He takes a long time. I stare at the stubbly back of his neck, his shoulders rising with breath. He must be remembering, for the shelves are not labeled, and neither are the glass bottles that crowd each shelf: some green, some brown, some purple, some with droppers for lids, some with burnt driftwood corks. I imagine he enjoys remembering, I imagine there is a story that explains each one, which ridge and what kind of coyote and how the rain came. These are Ed's blue-glass stories, these bottles and vials he considers—first this one and then that one, the glass clinking like small bells as he thumbs through them. I don't care how long this takes, how long he spends remembering. I'll wait.[17]

In this vignette, Wilkins introduces us to Ed Dempsey. But then Wilkins wanders into his own mind—and, theoretically at least, the mind of Dempsey. Time halts and we're left lingering in thought, at least until Dempsey's clinking glass bottles yanks Wilkins back to the present.

Along with speed of scene, writers can also use incident frequency to highlight how we misperceive time. Incident frequency is the number of times an incident occurs in the real world in comparison to the number of times it occurs on the page. Three categories exist.

Normative frequency is when the number of occurrences of a real event match the number of times the event is shown on the page. Almost all creative writers often use the normative frequency.

Iterative frequency is when the number of occurrences on the page is lower than the number of occurrences in the real world. The iterative is best used to remove nonnovel experiences. Writers don't need to talk about all the times they've gotten dressed for school. They need only mention those novel days (first or important days) getting dressed for school. Creative nonfiction often uses the iterative frequency to reduce nonnovel experiences.

Repetitive frequency is when the number of occurrences on the page exceeds the number of occurrences in the real world. Repetitive is best used to highlight fight or flight and novel experiences. Writers use repetitive frequency to highlight how an issue is so powerful we keep returning to it.

An example of a repetitive incident is from Tim O'Brien's *The Things They Carry.* In his story, "How to Tell a True War Story," the narrator keeps returning, again and again, to how one of his fellow soldiers dies. The narrator keeps repeating the story because he is trying to exactly capture how to tell this one death moment. Each time the narrator fails. So he tries again. The narrator says:

> This one does it for me. I've told it before—many times, many versions—but here's what actually happened.
>
> We crossed that river and marched west into the mountains. On the third day, Curt Lemon stepped on a booby-trapped 105 round. He was playing catch with Rat Kiley, laughing, and then he was dead.[18]

A few pages later, the narrator says:

> This one wakes me up.
>
> In the mountains that day, I watched Lemon turn sideways. He laughed and said something to Rat Kiley. Then he took a peculiar half step, moving from shade into bright sunlight, and the booby-trapped 105 blew him into a tree.[19]

And a few more pages later, the narrator says:

> Twenty years later, I can still see the sunlight on Lemon's face. I can still see him turning, looking back at Rat Kiley, then he laughed

> and took that curious half step from shade into sunlight, his face suddenly brown and shining, and when his foot touched down, in that instant, he must've thought it was the sunlight that was killing him. It was not the sunlight. It was a rigged 105 round. But if I could ever get the story right, how the sun seemed to gather around him and pick him up and lift him high into a tree, if I could somehow re-create the fatal whiteness of that light, the quick glare, the obvious cause and effect, then you would believe the last thing Curt Lemon believed, which for him must've been the final truth.[20]

And maybe that is the key with repetitive incidents. Repetitive frequency is rarely used, but it is best used when we need to retell a story over and over again or tell it in new ways as we, the narrators of our own life, search for our final truths while knowing that they might never be achievable.

In the end, creative writers should remember that time is vital, but, still, we misinterpret how long important and mundane events take. It is this misinterpreting that might be most powerful to our writing because once we understand how and why time is skewed, we can use that skewing to benefit our writing. We can focus on those moments that affect our perception of time so that we affect the reader and how they perceive time in their minds and on the page.

If writers can make readers feel as if time is being summarized and gapped until it might never exist during the quiet moments of our lives, we give them an authentic reality. If we pause or dilate scenes until time stretches nearly forever or repeat a scene again and again, when we live those novel moments, then we allow readers to experience the warped, essential experiences that compel us to write in the first place, to create for our readers an experience of time that mirrors the dramatic elongation of time, especially during those Bengal tiger moments of our lives.

3

Sipping from the Transmitter: Theorizing the "Potential Essay"

Lawrence Lenhart

The most surprising word I've ever encountered was my son's first. After months of sputtering, babbling, incanting—electric with prosody—he came out with it: a word. I won't tell you that word here, now, because maybe it was a secret. Later, when I replayed the syllables—even the imagining can activate the temporal lobe—I was proud of his inflection.

A list of other surprising words:

lighght
s ee k
Chikin
deathfugue
wordwall
breathturn

icethorn
redgold
silverflamed
heavenmetal
sd
yr
supercalifragilisticexpialidocious
bababadalgharaghtakamminarronnkonnbronntonnerronntuonnt
hunntrovarrhounawnskawntoohoohoordenenthurnuk

Some words cannot be taken out of context because they are their own context: brands, surnames, memes, or slurs. At a high-enough frequency, a word, any word, can signal alarm. When whispered, the same word becomes confidential. If repeated: neurotic. In 1968, a word provoked outrage when senators warned Americans that their tax dollars were being frittered away on concrete poetry. Aram Sorayan's one-word poem, "lighght" (that's it, the whole poem), was purchased for *The American Literary Anthology* by the National Endowment for the Arts for $750 (or the equivalent of $5,500 today). Sorayan has said the poem tries to make "the ineffable, which is light—which we only know about because it illuminates something else—into a *thing*. An extra 'gh' does it. . . . It's sculptural on that level." Whether or not "lighght" accomplishes what it sets out to do, I pause every time I see it—on a computer screen, the printed page, a whiteboard, or tattooed on an arm. I get this incredible urge to take out my wallet and throw money at it.

In-demand freelance writers may charge up to $0.30 for each and every word they type. At this rate, my last sentence could fetch $4.50 ($4.80 if you count the hyphenated compound as two). Idiomatically, this practice is centuries-old with two-bit, four-bit, ten-dollar, and hundred-dollar words connoting highfalutin diction. Notice how inflation reappraises the idiom. But the biological value of a word—its neurophysiological worth—has remained constant over time.

"Words are ancillary to content." So says the epigraph to N. H. Pritchard's *Matrix*. Sometimes I think the dramaturgy of diction is the only thing worth fighting for. Like how so many of my sentences sound like an argument with myself, a competition: replacing a word with its better, then its best, recalibrating my diction until I've been accused (just once) of impersonating someone with Tourette's. It's just how I talk. How to explain that each sentence is a predicament, and I must riff my way back into silence.

I can remember the way my mind boggled when, at Diné College, Orlando White's poetry students showed me the ambigrams they'd been working on. Cognitive scientist Douglas R. Hofstadter defines ambigrams like this: "Calligraphic design that manages to squeeze two different readings into the selfsame set of curves." The gesture transforms the word into a wheel. Turning the paper does not distort its legibility. With its constant morphology, the word seems visually pure. Rendered like this, a word has infinite potential.

"But a word is a bottomless pit," Lyn Hejinian says in *My Life*. It's true that if you take these pages out from beneath your nose, you're probably looking toward crust, mantle, core. Imagine that core, and keep going. That's how deep I beg my diction to go. Words are my favorite sustainable resource. Because they are bottomless, you can extract from them bottomlessly.

Before that first word, my son used to say, "Gauww?!" as he pointed at objects. Every utterance was its own vector. "Book," I obliged. "Gauww?!" "Soap dish." "Gauww?!" "Cat." "Chewing gum." "Asparagus." "Gus!" he estimated. "Good," I said. "That's pretty good."

When I started college, I only knew that I loved words. I wanted to major in them, one way or another. Before committing to "writer," I first tried electroencephalographic (EEG) technician. In a fusty basement in

Pittsburgh—five stories beneath the swivel chairs of cognitive-linguistic professors—I smeared dozens of electrodes with conductive adhesive gel. I fit the net of rubber nubbins over the participants' brain regions. After situating the central electrodes, I tugged the headband until the frontal and prefrontal terminals were in place. I shimmied it over the temporal, the parietal, and with one last yank, the occipital. Just like that, the whole scalp was covered by small cylinders that resemble the nozzles fitted to the end of sprinklers. After some brief instructions in the sound-proofed booth—"Here's how to respond," "Here's how to pause," and "Don't fall asleep, really, don't fall asleep, you won't get paid,"—I left the participant utterly alone for the most monotonous ninety minutes of their life. I sat alone on the other side of that booth in the drop-ceiling basement staring at an oversized computer monitor. The only sign of life was the squiggling of brainwaves throughout the test, all carefully time-locked to the display of specific words within sentence after sentence after sentence. This meant I could observe the participant's reaction to each and every word. Watching the neural oscillations, I reimagined that slogan from the Partnership for a Drug-Free America: "This is your brain on words."

When humans encounter words they don't know, they have an involuntary reaction. Neurophysiologist Elvira Khachatryan calls it the "Wait, what?" moment of language processing. Technically, the cognitive-linguistic paradigm associated with "Wait, what?" is called the N400.

Electrophysiologically speaking, when we encounter a word, ions flow in the brain, pushing and pulling the electrons to the surface. It's the quantitative difference between the push and pull (fluctuations in voltage) that generates "potentials." Every time we read a word, there is a weak voltage on the surface of our scalp. I fantasize about someday creating a word that might create an exit wound, surging through bone, pericranium, connective tissue, aponeurotic layer, more connective tissue, and skin. Surely, James Joyce's "onomatopoeia" for the thunderclap accentuating Adam and Eve's Fall ("bababadalghar aghtakamminarronnkonnbronntonnerronntuonnthunntrovarrho unawnskawntoohoohoordenenthurnuk") was meant to summon this kind of force.

When a potential is measured as a direct result of a cognitive event like language processing, it's known as an event-related potential (ERP). These events transpire in less than a second.

Come to think of it: How would an electrophysiologist even measure the millisecond response to a word like *bababadalgharag htakamminarronnkonnbronntonnerronntuonnthunntrovarrhoun awnskawntoohoohoordenenthurnuk*, considering it takes several seconds to read?

In Len Platt's introduction to *Finnegans Wake*, there is the admission that the book is "virtually unreadable" (a heck of a way to introduce a book someone just spent their hard-earned money to acquire), "not because it has no meaning but, on the contrary, because it allows for such potentiality of meaning—to the extent that some readers have claimed it can mean anything and everything." Platt goes on to point out this is "patently untrue," since the reader can't transform the book into, say, *Moby Dick*—no matter how inventive their interpretation.

While they took the test, I watched the participants' brainwaves cresting and troughing on the monitor. I was on the lookout for unexpected noise or idle electrodes or delta waves. I'd knock on the booth, and the door would swing open. "I'm just here to wake you up," I'd say. "I wasn't sleeping," they often lied. I wanted to challenge them. "I have some data here that says otherwise." Once they left, I'd slosh together a saline solution to clean the electrode net and then save the data before running it through statistical software, which filtered the noise and isolated ERPs like the N100, P200, P300, P600, and especially the N400.

When I read a sentence, there are phantom words, invisibly inked, between the words: alternate linguistic realities—a subconscious matrix of expectations, associations, and aberrations. The phantom words flicker for a few hundred milliseconds before disintegrating. To read like a writer, one must hallucinate even the words that weren't chosen. In the blank slur that is the *interword*, I sense a shortlist of words averted. Like the late comic Mitch Hedberg's joke about "Do Not Disturb" signs, we process and valuate individual words faster than we do sentences.[1] And while it's unlikely that the hotel sign manufacturer intends for polysemy (Greek for "many signs"), the L=A=N=G=U=A=G=E poets of the early 1970s made it their solemn mission. They wrote in a manner that maximized deduction at the paragraph level while minimizing deduction at the sentence level. By "torqueing" sentence structure through unconventional syntax, the authors amplified polysemy, always training the reader's attention on the activity of language, not its meaning.

When I ask my parents what my first word was, my mom says: "Mom." Dad says, "I'm still waiting to hear it." It's his clever way of saying we don't speak the same language, have the same interests, have occasion to talk much at all.

The theory of the "new sentence" (Ron Silliman's now forty-year-old term for it) was really just a recapitulation of the two most influential Steins from the first half of the twentieth century. From Ludwig Wittgenstein's *Philosophical Investigations*: "When I say that the orders 'Bring me sugar' and 'Bring me milk' make sense, but not the combination 'Milk me sugar,' that does not mean that the utterance of this combination has no effect."[2] And Gertrude Stein in *How to Write*: "Within itself. A part of a sentence may be sentence without their meaning." Silliman, failing to find any redemptive theories of the sentence from an entire century of linguists, especially as it might apply to literature, turned to these two, as well as his contemporaries.

In "Writing (as) (and) thinking," Hélène Aji notes that Charles Bernstein (a friend of Silliman's) "calls attention to the combinatory nature of language and to the potentialities of deviant combinations." An equally clunky version goes like this: "Poetry is a graphic form of unrighting the publicly codified colocation of a grapheme with symbolized ordinary writing and speech usage and the imaging function of the mind." Likewise, Aji directs us to the automaticity of Lyn Hejinian's process: "The language itself materializes thought; the writing realizes ideas. One discovers what one thinks, sees, says, and as the words unfold in the work, the work, directed by form, extends outward." Elsewhere, Hejinian is charmingly coy: "I am suddenly aware that phrases happen."

First discovered in 1980, the N400 is, (very) technically speaking, a "large negativity with a broad (parietally maximal) scalp distribution, peaking around 400 ms (largest for semantic anomalies, but also present for improbable but sensible endings)"—this according to Marta Kutas, one of the neuroscientists who first discovered the linguistically significant paradigm. In other words—because when re-presenting science, we often need them—when readers encounter a sentence with unexpected diction, their brains react following a certain electrophysiological pattern. In the original article in *Science* (1980), "Reading Senseless Sentences: Brain Potentials Reflect Semantic Incongruity," Kutas and Hillyard use these compositional examples to demonstrate semantic congruity, moderate semantic incongruity, and strong semantic incongruity, respectively. Expected: He took a sip from the *fountain*. Somewhat unexpected: He took a sip from the *waterfall*. Not at all expected: He took a sip from the *transmitter*. Four hundred milliseconds after the reader encounters a misfit word like *transmitter* (or about the amount of time it takes the average person to blink), the entire scalp buzzes with electrical activity, most noticeably at the parietal lobe where sensory and language processing occurs. The stronger the incongruity, the greater the amplitude. This amplitude, this time-locked peak, is language crashing against the brain.

Reading has never been the same since working in that EEG lab. That year, I took to slinking my fingers through my longish hair while reading. I traced each strand to its follicle, thinking I might feel the weak shock that is "word processing" on my fingertip. It was that watershed year when the global number of text messages officially surpassed the number of phone calls. My long-distance girlfriend wasn't messing around anymore; we pooled our funds and treated ourselves to matching "unlimited texting" plans. Words buzzed into my life, and I found myself reading in places I had never associated with language: elevators, urinals, crosswalks, and queues. Even then, when I read her texts, it was happening: each word resulted in a faint electrical flutter on the scalp so that when she was brainstorming via text about her Halloween costume (something about Bohemian discotheque), I was cognizant of an electrical brainstorm of my own. That semester, I read lines of William Carlos Williams, wondering what kind of parietal peaks it must be inducing. When my poetry professor, Ben Lerner, heard about the EEG lab where I was working, he suggested that rather than quizzes, he should just have students submit their brainwave activity from their reading. Enough N400s, and you get an A. Delta waves (sleep), and you fail.

Of course, researchers didn't have creative writing in mind when they first discovered the N400. Rather, they were interested in the potential for the N400 to explain and mitigate reading impairment and language disorders. In the thirty years since the discovery, Kutas and Federmeier describe applications in "language processing; object, face, action, and gesture processing; mathematical cognition; semantic and recognition memory; and a variety of developmental and acquired disorders." In short, the N400 is one of the premiere diagnostic tools for all manner of linguistic conditions.

In his introduction to *The Next American Essay,* John D'Agata writes, "A fetus at eight weeks has developed its ears but not yet the ability to hear. What this means is that anything you read to a fetus will go in one ear, but not come out." And yet, my son's first word was one of the first I ever read to him, prenatally. "[Word]," I said to him. "[Word]," he said back.

My wife, who has never owned a dog, says we can get one as soon as our son asks the question. For a few days, I try to train him to say *dog*. I prolong the phonemes: *duh-aww-guh duh-aww-guh duh-aww-guh daw-guh daw-guh dawg dog.* If only he says it with a pleading look. Like he says, "Gauww?!" It reminds me of Donald Barthelme's story, "Chablis," in which a mother tells a father their baby wants a Cairn Terrier. She knows this telepathically. No language required. My wife makes a rule, though: it must be a complete grammatical sentence.

I often left the lab as it was getting dark, jaywalking toward the secular Cathedral of Learning where my literature class was held. My favorite book that semester was Jonathan Safran Foer's *Everything Is Illuminated.* The secondary character Alexander Perchov is a Ukrainian translator who is notorious for his unchecked use of the thesaurus. Perchov's trials in translation resulted in many surprising phrases. "Pygmy allowance" and "seeing-eye bitch" are just a few of the surprising deviations from the high(er)-cloze-probability words *small* allowance and seeing-eye *dog*, respectively. After class, I wished I could return to the lab, read the playful book while hooked up to an EEG—to see the way an unexpected word jolted me as a reader, forced my cognitive-linguistic sensibilities to be destabilized for a half-second while I found ways to accommodate the dictional aberrations. Instead, I resorted to annoying my girlfriend by reading my favorite paragraphs (sometimes pages!) over the phone. I hear myself now, mansplaining these passages for tens of minutes, foolishly trying to understand an effect that lasted for halves of seconds.

That semester was the first time I heard the word "experimental" outside of a science class. Each time Lerner used it to us—introductory poetry students—it made me skeptical of Creeley, of Spahr, of Williams, and of Rankine. I looked at the sentences, wondering what the poet's hypothesis had been, and whether the syntactic trajectory supported that hypothesis or refuted it. I regarded the clauses as variables, and each word as datum.[3]

With a surname like D'Agata, it's no wonder how the foremost writer of the lyric essay arrived at his genre: look at all the schwas. But let's not forget how ambivalently he titled the oft-cited introduction to *Seneca Review*'s lyric anthology: "We Might as Well Call it the Lyric Essay." Resistant to the nomenclature, D'Agata writes: "'lyric essay' is no less an example of lipstick on a pig—which I think is why you'll find that is has fallen out of favor with a lot of the writers in this book." The lyric essay is an oxymoron, "an essay that's also a lyric; a kind of logic that wants to sing; an argument that has no chance of proving anything." It is a prose that poems, an essay that lyres.

If you look at the N400 on a coordinate grid, you will find it resembles Freytag's Pyramid. There is a steep incline as it progresses along the x-axis, peaking/climaxing before its attenuation. There is a story behind every N400, some causal concealment. "Tell me the story about the time you took a sip from the transmitter," I beg my students. "Tell me about the time you unlocked a plant, staggered a cloud, prayed a guitar." There's always one per classroom who won't accept the challenge, who thinks it's just stupid. The others, giddy to negotiate their contract with language and memory, are willing to rappel into that bottomless pit.

A decade has passed since the fusty basement. I am reminded by a researcher halfway across the world that all words, not just the "incongruous" ones, elicit an N400, however faint. Even the most straightforward sentence—"I" "want" "a" "dog," for instance—becomes a site of cognitive trepidation. Each word potentiates the ones surrounding it. Each sentence is the site of a linguistic unnerving.

There is a difference between "inventing" a new language and a new *kind* of language. Anthony Burgess, who taught courses in phonetics and literature, developed his own Anglo-Russian teen slang for *A Clockwork Orange*:

> Oh it was gorgeousness and georgeosity made flesh. The trombones crunched redgold under by bed, and behind my gulliver the trumpets three-wise silverflamed, and there by the door the timps rolling through my guts and out again crunched like candy thunder. Oh, it was wonder of wonders. And then, a bird of like rarest spun heavenmetal, or like silvery wine flowing in a spaceship, gravity all nonsense now, came the violin solo above all the other strings, and those strings were like a cage of silk round my bed. Then flute and oboe bored, like worms of like platinum, into the thick thick toffee gold and silver. I was in such bliss, my brothers.

For the most part, the paragraph looks like madman play mad libs with an otherwise coherent English paragraph. A quick surgery of nouns and verbs can restore the semblance of traditional meaning. This uninhibited flex of language can be found in Cathy Park Hong's poetry as well. In his review of *Dance Dance Revolution*, John Yau points out that the speaker (known as "The Guide") "speaks a lingua franca or what she calls 'Desert Creole,' a cacophonous mishmash of puns, pidgin languages, malapropisms, neologisms, and portmanteaus." The language is "an amalgam of some three hundred languages and dialects."

But these languages, coded or globalized, are altogether different than the language system in *Finnegans Wake.* As Platt describes it, *Wake* "announces a new 'revolution of the word'. . . one that works not to stabilize the world, but, rather, to unfix it in a wild diversity of possible or potential significance." Whether it's virtuosic or incoherent,[4] it is a new *kind* of language.

To call a work "potentially significant" makes that work sound unread, un-understood, or simply undealt with. By and large, we have not met the challenges associated with Joyce's revolution; if we had, scholars would not be publishing articles titled "*Finnegans Wake* for Dummies" in the flagship journal associated with Joycean studies.

Potential energy is *stored* energy. It is the book willing, able, to do its work. It is the word, resting on the page until perceived. The transfer from potential to kinetic energy requires work, in this case, cognitive work: the kind that results in firing signals or "spikes." Consider the work a word requires. Word. Spike. Consider the work two words require. Word, word. Spike, spike. Now three. Now four. Now sixty thousand. A spike train is, according to neuroscientist Devika Garg, a "combinatorial sequence of spikes and silences"—a kind of binary code in which the zeros represent the latency period between stimulus onset and response.

Imagine reading *Wake* word by weird word in sequence like a slow-drip IV. Watch as the words, their wild semantic potentials, cause ERPs. Aggregate and analyze the massive data stream.

Literary scholar Franco Moretti, sometimes referred to as the Linnaeus of literary theory (elsewhere Galileo or Darwin), has proposed a strategy known as "distant reading." *The New York Times* provocatively summarizes Moretti's scholarship like this: "To understand literature . . . we must stop reading books." His hypothesis-testing, computational modeling, and quantitative analysis supplants the human perceiver of literature. Think concordances and word clouds, vectors, and plot diagrams. In Moretti's world, the "protagonist" is "the character [who minimizes] the sum of the distances to all other vertices." It can be intuited—along with antagonist, genre, subplot, theme, you name it—from grammatical and semantic signals (yes, words) alone. *The New York Times* again:

> People recognize, say, Gothic literature based on castles, revenants, brooding atmospheres, and the greater frequency of words like "tremble" and "ruin." Computers recognize Gothic literature based on the greater frequency of words like . . . "the."

But a regimen of distant reading cannot terminate the author's relationship with her first reader: herself.

Rikki Ducornet, whom I've often called the best word-for-word sentence writer in the world (to borrow from the parlance of boxing), says: "Writing *is* reading and reading a way back to the initial impulse." A Ducornet sentence is often visual and discursive, set off by em dashes, parentheticals, commas, italics, capitonyms; each small choice is a bonfire of meaning. My favorite sentence in *The Deep Zoo* goes like this: "In other words, to write in the light of childhood's burning alcohol, with the irresistible ink of tigers and the cautious uncaging of our own Deep Zoo, we need to be attentive and fearless—above all very curious—and all at the same time."

At bedtime, my son's nursery feels more like linguistics lab. With just one simple or compound sentence per page, a children's book follows the experimental control rules of the EEG lab. Words like "room," "telephone," "balloon," "cow," "moon," "bears," "chairs," "kittens," "mittens," "mush," and "hush" all conspire to mean something to my son. I imagine them the way they must mean to my son, all mashed up—"In the great green balloon, there was a bear and a red cow and a picture of mush jumping over the telephone."—until suddenly they're not. Each page is a "Gauww?" waiting to happen. It's his own way of saying, "Wait, what?" It's the N400 as bedtime ritual.

Don't read books, just their words. To comprehend our electric-creative potential, we must be willing to encounter words in a singular way—as they lie in the rented space of the sentence, begging their way into our blinking brains.

4

The Brain Is a Master Class

Dave Madden

Robert Atwan once wrote, "What essays give you is a mind at work," and I used to bandy about this quotation as a writer of essays, and as a teacher of them, because for a time I thought it was useful. I thought it pointed to what was most glorious about the essay: its ability to show off some first-rate creative thinking. Then I moved to San Francisco, and in the bar of the hotel we stayed in while looking for a home, I saw hung above the shelves of underlit bottles a multi-bulbed sign that read Be Amazing. Why, I wondered, was this the imperative of our time? Magicians are amazing. To amaze someone is to dazzle her, or to leave her stunned by you. And it seemed to me that if our goal is stunning other people with ourselves, we're not hearing other people or even really talking to them. I write essays to communicate, and I read them to engage in someone else's new ideas. And that, I realized, was key: engagement between writer and reader. There's an absorptive, maybe even interactive quality to the essay, which Atwan's bit doesn't address.

Then I found another quotation: “The rhetorical function of essaying is not merely to transmit the essayist’s thoughts but to convey the feeling of their movement and thereby to induce an experience of thought in the reader.” Who wrote it was a guy named R. Lane Kauffmann, a professor in what was then called the Hispanic studies department at Rice. Not the sort of person you’d tend to find in the craft books, but look at what he’s pointing to: inducing the experience of thought in the reader. Here’s where the absorption lies. It’s not enough to show your mind at work. We essayists need to get our readers’ minds working on their own.

* * *

Two obvious questions:

1. What does a mind at work even look like?
2. How can we, *après* Kauffmann, convey the feeling of thoughts’ movement?

To answer these questions I went looking for books on the brain’s structures and biological functions, and I read the work of cognitive scientists who studied human perception and theories of learning—how the mind works. I learned some things I think can help us write essays that engage our readers more fully, but, first, I need to sell you on an argument you may not want to buy. *Essayist : reader :: teacher : student* may be the sort of analogy only a professor of creative writing would come up with, but even at their most personal, essays feel didactic. We’re often teaching our readers how we want to be understood. One way, then, to identify “good writing” is by how well it gets readers to think and feel *alongside* the essayist. Again, it comes back to absorption. I want my readers to fall into the induced experience of thought the way I fall, when reading, into the worlds of my favorite novels so fully I forget I'm just looking at words on a page. Absorption in fiction—ditto narrative nonfiction—is created through sensory detail. But how is it created in an essay, like this one, that doesn't have a story to tell?

* * *

More than 160 billion cells make up the brain. (The heart, by comparison, has a measly 2 billion.) The cells worth our time in this essay are called neurons, and they look like those spiked iron balls certain foreboding videogame bosses swing menacingly over their heads at you. The ball of the flail contains the cell nucleus, and its spikes are called dendrites; they receive electrochemical signals from other neurons. Neurons send their signals from the nucleus down a chain called the axon, but there's not going to be a test, here, so ignore the vocab for the process. Stimuli—sounds, smells, etc.—fire our neurons, and when two neurons fire together, the connections between them strengthen. "Neurons that fire together wire together" goes the neuroscientific maxim, and it's the biological basis for learning. (And not just human learning: Ivan Pavlov fired "I hear a buzzer" neurons the same time he fired "I smell food" neurons to the point where the "I smell food" neurons didn't need to fire anymore for the "Let's drool now" neurons to do so.)

What I mean by "fire" involves the swapping of potassium and sodium ions (among others) in a physiological process we don't have time to explore in full here, but one thing it will help to shine a light on is that this firing/swapping occurs at the synapse: a microscopic bit of space between the axon and dendrite. Which is to say: signals leap. Such leaps—as you read this, as you take in passing audio information, as you breathe, blink, fart, and daydream—are continually happening across all the brain's parts.

So, what does a mind at work look like? It's all quickness and light.

* * *

If I give you a list of people to remember and a woman named Baker is on the list, you're less likely to remember her later than you would a woman who *is* a baker. The reason behind this is harsh: language barely signifies. Baker, the name, fires neurons associated with naming and language. That's it. But baker, the occupation, fires not only language neurons but those associated with smell, food, memories set in bakeries, and so on. Anything encoded semantically—that is in terms of its significance and meaning—will have a better chance of making it to long-term memory storage than things encoded only visually or acoustically.

It's called the *Baker-baker* paradox and it shows us a way to write better: encode your stuff semantically. Generate layers of associative meaning. My favorite example of this comes from the "Findings" column in *Harper's*—the magazine's back-of-the-book assemblage of the month's findings from the scientific community. In essence, it's a prose poem built out of facts, but these facts are always re-engineered to convey both information *and* emotion. Once, I sat down to talk with its author, Rafil Kroll-Zaidi, about his process. He pointed me to a certain finding from years before we met:

A cat gave birth to a dog in Brazil.

"First problem with that," he told me, "is what's interesting is that a cat gave birth to a dog. 'In Brazil' is just dead language sitting at the end of the sentence. You don't care about it by the time it happens." He suggested a revision:

In Brazil, a cat gave birth to a dog.

"So you've got the kicker at the end," he said. Then he gave me a funny look. "Or. . .? Better. . .? Can you guess?"

I couldn't guess. I waited for him to tell me.

A Brazilian cat gave birth to a dog.

Brazilian shifts from adverbial information to an adjective, he explained, and when that happens, something goes queer about the sentence. The sentence is already confusing: A cat shouldn't have given birth to a dog. Kroll-Zaidi delivers the finding in such a way to amplify that confusion, thus making it far more memorable and affecting than, say, "According to a 2006 Reuters report, a woman claimed 'her cat Mimi had given birth to . . . three puppies as well as three kittens.'" (Whether you're OK with overlooking the additional two dogs is another essay for another time.)

Here, on the page, was the quickness and light I saw going on in the brain, those rapid associations engineered through language. Kroll-Zaidi is looking to build layers of affect: information and mood, punful disturbances. It's a lot like what makes Julie Smith, David Foster Wallace's protagonist in the short story "Little Expressionless Animals," so good at *Jeopardy!* Smith spends much of her childhood reading and consolidating to her long-term memory "an obscure and limited-edition Canadian encyclopedia called *LaPlace's Guide to Total Data*," which we're meant to believe is one reason she becomes the first-ever *Jeopardy!* contestant to stay on as champion for more than the regulated five-day run. (The story is from the 1980s, long before the rule changed, allowing for serial winner Ken Jennings et al.) Smith wins 740 games in a row. One other reason for her success is more occult. "This girl not only kicks facts in the ass," is how one character puts it. "This girl informs trivia with import. She makes it human, something with the power to emote, evoke, induce, cathart."

There it is again: *induce*. Semantic encoding gives us the power to induce an experience in the reader.

* * *

About ten years ago—to give you a sense of how new these findings are—Robert Bjork, a cognitive scientist at UCLA, ran an experiment on learning. He set up two classes of students, all of whom had to learn about the work of twelve artists. The students in Class A "block sorted" the work, studying six different paintings by one artist before moving onto the next. In Class B, they "interleaved" the work, studying one painting by one artist at a time and cycling through the twelve artists in six rounds. All students saw the same seventy-two paintings. Then Bjork tested both classes by asking them to identify which of the twelve artists created a painting they hadn't seen before.

Guess which class did a better job on the test.

Class B, with the interleaved study, did a better job. Findings like this led Bjork to coin the term "desirable difficulties": conditions that appear to impede performance during training but which actually lead to better long-term retention and retrieval of the material. Researchers found all sorts of other such "difficulties." My favorite example is the study in which one class studied materials on a fictional animal printed in good old, boring Arial font and the other class studied the same materials printed in Comic Sans MS at 60% grayscale. Again, class B did better on tests.

Does this mean essayists should submit our manuscripts printed in Comic Sans? You're welcome to try. I won't, but as a writer I'm interested in this notion of creating desirable difficulties, and more specifically in interleaving. Interleaving looks a lot like what we in the nonfiction community call a "braided essay," in which two or more topics are woven together to form an essay's throughline.

(John McPhee's "The Search for Marvin Gardens" and Jo Ann Beard's "Coyotes" are good examples.) Bjork's findings suggest the braided essay might be more than just a nice way to shape a piece; the structure might actually help a reader engage more deeply with an essay and retain more information from it.

And it's nice to think of difficulty as being desirable. So often, in my first drafts I worry over what readers will understand or how much they'll follow me on a line of thought. Or I'll get a wild idea to leap from beer cans, say, to the shoes of Imelda Marcos, and I'll worry that readers won't follow me there.

But readers will follow you there, and they'll be happier for making the jump. More than "show don't tell," the phrase I repeat *ad nauseam* in my classrooms is "leap don't creep." Make wild associations and connections without worrying about spelling everything out. Joy Castro, one of the finest memoirists I know, told me recently always to assume my reader is smarter than I am. "Leap don't creep" is not only a reminder to write up to such readers, giving them desirable difficulties to engage with, but it also makes me write in a way that mirrors the working of their active, brilliant minds.

* * *

The brain is a master class on how to write nonfiction. The better we understand its processes and its landscape, the more artful our writing can be. It's like sending fiction students out in the world to capture the way light falls on a tree or to eavesdrop on conversations so as to understand dialogue. When you write to build a world, go out into the world. But when you write to induce a mind to think, why not go, *Poseidon Adventure*-like, into the mind?

5

Brain on Fire

Nicole Walker

My father grew up on Dearborn Street, in Salt Lake City, Utah, but after his dad died, his mom moved to McClelland Street. In my memory, I conflated Dearborn the Street with Dearborn the town in Detroit and imagined my dad sitting on the stoop of a brownstone playing kick the can down the center of his Detroit-centric home. On McClelland, my grandmother gave us big wooden, brick-colored blocks to play on the carpet with our cousins while she fed us apples sprinkled with salt. I fell off my tricycle while riding around the block around her street. She fell down the stairs while babysitting me in her McClelland house after drinking too much and had to move to a one-story condo. Or, at least that's the connection I made. My dad wasn't big on nostalgia and in so not telling us stories, the stories slip in. In my mind, my dad had deep city roots. The sound of Grandma's ribs hitting stairs is a crack—a shebang, a lightning bolt. It keeps all the brick, brownstone images together. Now, if I write a story about my grandmother or my father, it goes brown, stone, brick, crack.

This may be a way to consider how we make essays that generate emotional and empathetic response. Argument and narrative are great for hooking people into an essay, but how do we make it resonate within the reader beyond intellectual understanding? When we teach creative writing, we ask our students to be specific, to include idiosyncratic detail. We know we like to read details, we know it makes the writing more mimetic to life, but we also understand, subconsciously, that the details make something bigger than the sum of the parts and that the details exist to create visceral opportunities for emotional and empathetic connection.

My husband, Erik, spent his elementary school years on 14th Street but then his family moved to Wellington, which is just down the street from the same Dearborn Street my dad grew up on. For the first time in my life, because I was dating my future husband, I drove by that street. Erik pointed out to me the elementary school on Dearborn. My mother told me—she who would rather fill in the gaps than leave it to my collection of misremembered memories—that my aunt was impregnated by the janitor of that elementary school. I click through the names in my memory: Greg, no that's whom my cousin married when she was seventeen. Jimmy, no that's my aunt's second husband. Randy? There were two Randys. The first was my aunt's hippy boyfriend who drove a Trans Am and who took us to a hotel room at Snowbird Ski Area with a gas fireplace. I put my doll too close to the fire. Her hair melted to the back of her head. The other Randy was a real asshole who married my other aunt and took all her money. What was the janitor's name? My grandma, my dad's mom, like my dad himself, wouldn't speak it so now the idea of him just tumbles into the gap, and I picture the janitor with a mop and a mustache and my aunt's bulbous belly, carrying my cousin in her body. My cousin is only six months younger than I even though my dad was eight years older than my aunts. We played with grandma's wooden blocks together. We built whole brownstones out of them.

Functional MRIs have lately been able to show how visual images make an impact on the brain. As a writer, I want to know how many triggers I can pull using images, word choice, and metaphors to make the associations across the brain chime together in the way they do for me: brick to janitor to mop to brick. When I write nonfiction, I think of things like argument, memory, and story, but what I rely on, since I usually talk myself out of my argument, since my memory is a steel sieve, since story reads like interrupted anecdote, is image. I use image like a leitmotif that supposedly holds the story together. But does it? Do the triggering images trigger the associations I mean them to in the reader?

When my boyfriend Alex punched me in the jaw after he'd moved back into the house on Cora, he didn't mean it. He was drunk. I probably was drunk too. I don't remember exactly why he hit me. I was nagging or complaining or aggravating. This is not a domestic violence narrative. I bring it up only because it's the only time I remember being hit or even seeing anyone hit. My nose wasn't bloodied. I had lockjaw anyway. I do not begrudge him for this at all. What I begrudge him for is the weight I lost and then gained back after he broke up with me days/weeks/months after our fight. The white dress, size 3, I could wear for those months we were broken up and I was antagonizing him by coming to his apartment where the walls were painted a whiter white than my dress and we listened to so much Paul Simon that I still sing the lyrics "take this child far to Tucson Arizona, give her the wings to fly so she won't bother you no more" every time I think of that apartment on Hawthorne, before he moved in with me on Cora, and how I was such a bother to that poor guy that year he broke up with me and how I ate only one bite of hamburger from the McMenamins just down the street from his apartment.

Images can provoke specific synaptic activity. The connection between video game violence and real violence has been a political ball bounced between the right and the left to argue causes for violent adolescent behavior. According to research, there actually is some fundamental connection between violent images triggering violent centers of the brain. Joy Hirsch, Director of Columbia University's Center of Radiology, and her team, "using mainstream Hollywood movies, collected clips on three categories. One group depicted acts of violence. A second featured images of fear—including facial expressions of terror—but no actual violence. A third was composed of intense physical activity that did not involve the threat of violence, such as dance or sporting events."

Functional MRIs showed that the amygdala, which is the place where the fight or flight response resides, was triggered by those who watched the violent events but not by those who looked at images of reactions to violence or to dance or sporting events. As our optical nerve, which connects the eye to the brain, translates and sends signals to the brain about two-dimensional, single-sensory information from a video game or Hollywood movie, the brain twists and turns, trying to make sense of it. Why fight or flee an image? It's just a picture. But our brain says no, it's not just a picture. Adrenaline rushes. The word "just" disappears. Virtual images can stimulate a literal, physical response. The brain reacts to images in the same way it reacts to real-life experience.

Do words stimulate in the same way as images? First, we must figure out what we mean by images. W. J. T Mitchell in his critical essay "What Is an Image?" asks, "What isn't an image?" "We speak of pictures, statues, optical illusions, maps, diagrams, dreams, hallucinations, spectacles, projections, poems, patterns, memories, and even ideas as images, and the sheer diversity of this list would seem to make any systematic, unified understanding impossible."[1] Mitchell recognizes that not all images perform the same way in the brain—diagrams may be less metaphorically representative than pictures, but that we can concede there is a "family tree" upon which these different types of images hang. Mitchell then categorizes images in a figure that holds graphic images (like a picture of a brick) on the left side and verbal images (like the letter B. R. I. C. K.) on the right with perception with optical, perpetual, and mental in between. Hirsch's video game experiment focused on visual images, but do words stimulate a similar physical response? Can we imagine that the words that we use to construct an image and the words themselves (perhaps as the brain free-associates with those individual words or categorizes them) function similarly? How can we, as writers, recognize the possibilities of words-as-image and the word itself from creating a mental world for the essay, story, or poem we write? What is it that we writers want our word-images to do inside our reader's brains?

Writers make verbal images to create sense and sensory responses. Research led by neuroscientist Maximilian Riesenhuber of Georgetown University Medical Center reveals that the brain recognizes words as images, like B.R.I.C.K. You may "see" a brick, but you may also see the letters that make up the word brick. A picture of a brick and a picture of the word "brick" are both interpreted by your brain as images: "As your eyes scan these words, your brain seems to derive their meaning instantaneously. How are we able to recognize and interpret marks on a page so rapidly? A small, new study confirms that a specialized brain area recognizes printed words as pictures rather than by their meaning."[2] For the purposes of this essay, I make the point that words convey meaning in an informational, translated from individual letters to dictionary definition, kind of way as well as a graphic image of the word itself, giving one-word multiple avenues of sense-making.

Whereas W. J. T Mitchell argued that words transform into images through perception and optics, Riesenhuber argues that words themselves transmit information similar to pictures. Riesenhuber studied nonsense words that readers came to "know" through sound and familiarity of their shape, suggesting that we make meaning out of the graphemes of words as much as the phonetic and signifier elements of words: "Because the nonsense words had no meaning, Riesenhuber deduced that our neurons must respond to words' orthography—how they look—rather than their meaning." Words literally are understood as images, not the meaning the words intend to signify but as graphical images themselves, by the brain. They transmit information through mental processes as immediate as pictures. This is interesting to writers because it complicates W. J. T Mitchell's hanging images—if words act as both picture and meaningful text, words may stimulate more than one area of the brain at once: the "picture image" receptors and the "meaning" receptors. If a writer's goal is to make as much "sense" as possible, writers can possibly corral images to spark multiple connections.

Neuroscientists continue to study how words stimulate mental activity across the brain. Princeton researchers used functional MRIs to discover how the brain understands words differently than pictures. Head researcher Matthew Botvinick says, "The thought is that there are many things that can be expressed with language that are more difficult to capture in a picture. Our study dealt with concrete objects, things that are easy to put into a picture, but even then there was an interesting difference between generating a picture of a chair and generating a list of words that a person associates with 'chair.'" Botvinick and his fellow researcher, Francisco Pereira, asked study participants to visualize words as they were presented to them. By using categorizing types of words, the researchers could, using functional MRIs, trace the areas of the brain that lit up. They were able to see that words like "chair," "table," "dresser," and "bed" all as one category: "At the same time, the team established all the words associated with 'furniture' by matching each fMRI image with related words from the Wikipedia-based list." But words like "chair" don't stick in one category like a graphic of a chair might. "Someone will start thinking of a chair and their mind wanders to the chair of a corporation then to Chairman Mao—you'd be surprised," Pereira said. "The brain tends to drift, with multiple processes taking place at the same time. If a person thinks about a table, then a lot of related words will come to mind, too. And we thought that if we want to understand what is in a person's mind when they think about anything concrete, we can follow those words."[3] Words have the capacity to drift. Homonyms, homophones, graphemes, and words as images stimulate areas across the brain, creating surprising connections.

Sucking up words as grapheme and words as signifier, the brain then categorizes. But just because the brain likes to categorize, it doesn't mean a word or a word-image has to be organized to only one category. Complex ideas expressed in language can light up several regions of the brain at once. We can trace this brain activity as the word "bed" might light up areas associated with "furniture" and "sex" and "sleep." As we write, then repeat certain images, put them in certain categories, we trace and retrace, making patterns. Our stories become a complicated light song.

I think of the communication between the aliens and the humans in *Close Encounters of the Third Kind*. I think of the brown bricks in my grandmother's closet in Salt Lake City that I would pull out and immediately be transported to another place—a category of Dearborn, Detroit, and brownstone that exists only in a real place in my mind. I put on an off-white dress and remember that one bite of hamburger at McMenamins and how desperately I wanted that moment to last forever—boyfriend coming back to me, thanks to my thin body and my off-white dress. He would let me into his white apartment now.

As we write, we may not know exactly what categories we trigger in our reader's brains but perhaps, like the light and music display in *Close Encounters*, by repeating different words, brownstone, Dearborn, janitor, perhaps we create our own category, pulling together an ever-tightening ball of "this particular story." Perhaps we pull connections across categories, making new associations possible through repetition, image, metaphor, and leitmotifs. Ian Sample, science editor for *The Guardian*, reports that "scientists have created an 'atlas of the brain' that reveals how the meanings of words are arranged across different regions of the organ."

"Like a colourful quilt laid over the cortex, the atlas displays in rainbow hues how individual words and the concepts they convey can be grouped together in clumps of white matter." Sample writes,

> No single brain region holds one word or concept. A single brain spot is associated with a number of related words. And each single word lights up many different brain spots. Together they make up networks that represent the meanings of each word we use: life and love; death and taxes; clouds, Florida and bra. All light up their own networks.[4]

Researchers Jack Gallant and Alexander Huth, using stories from the Moth Radio Hour, a public radio program featuring short, dramatic nonfiction pieces, recorded the subjects' brain activity as they performed their stories. Gallant and Huth discovered that there are, at least for these subjects, defined semantic pathways that show how specific words stimulate brain activity.

> The atlas shows how words and related terms exercise the same regions of the brain. For example, on the left-hand side of the brain, above the ear, is one of the tiny regions that represents the word "victim." The same region responds to "killed," "convicted,"

> "murdered" and "confessed." On the brain's right-hand side, near the top of the head, is one of the brain spots activated by family terms: "wife," "husband," "children," "parents."[5]

Across this atlas of brain activity, the word-images spark. The synapses ignite over here and then over there, the brain lights up. Isn't the writer's dream to turn the whole brain on fire?

In quantum physics, although I can't think of a billiard ball metaphorical equivalence, there's a theory called quantum entanglement that provides a good metaphor for how we can get multiple networks in the brain lit up simultaneously. "Quantum entanglement is a quantum mechanical phenomenon in which the quantum states of two or more objects have to be described with reference to each other, even though the individual objects may be spatially separated."[6] If you get one particle spinning in one network, its twin particle spins in another region. Think of a word-image, like bed, which might light up the "furniture" network as well as the "parent" network. By repeating the words "bed," "furniture," and "parent," you can make these different regions of the network glow. Then, by making that pattern, you can make the "furniture" a nostalgic, nurturing space as well as a place you put your quilt.

The Princeton study referenced above reaffirms this pattern making. Those word associations, lead author Pereira explained, can be thought of as "semantic threads" that can lead people to think of objects and concepts far from the original subject matter yet strangely related. "Someone will start thinking of a chair and their mind wanders to the chair of a corporation then to Chairman Mao—you'd be surprised," Pereira said. "The brain tends to drift, with multiple processes taking place at the same time. If a person thinks about a table, then a lot of related words will come to mind, too. And we thought that if we want to understand what is in a person's mind when they think about anything concrete, we can follow those words." These collections of images create new ways of thinking and new ways of showing how thought works. Now to stitch the images back together by returning to the quantum physics billiards metaphor: If we think of images as a cue ball and a pool stick and the brain is a billiard table covered in balls, pull back on the cue stick, write about the color white on the walls, the white of the dress, the whiteness of the boyfriend's face you forgot to mention, and then jam that stick ahead. Am I a good pool player or a crappy one? Did I hit the cue ball, make it roll to the right, make it bounce in the center of the solid blue which rolled into the center pocket while striking the solid green and making it roll into the left hole? Or am I a crappy writer and scratch the cue ball? Is this what I want from my narrative anyway? You can play different kinds of pool: Nine Ball, Eight Ball, One Pocket, Cutthroat, or 14.1 Continuous. All are played on a large, carpeted table with four corner and two side pockets, up to fifteen numbered and colored billiard balls, a cue stick and a cue ball. The object is to pocket a certain group of balls, sometimes in a particular order.

There isn't one way to write an essay. There isn't only one trajectory we must follow. But like billiards where players pocket a certain group of balls, writers group a certain collection of words. By paying particular attention to those words, by repeating them and making metaphors of them, we create a world of images, an image set, that houses our stories and allows them to cohere beyond the more common understanding of coherence. Through word-image gathering, the writing hangs together not through narrative or argument or other kinds of rhetoric but through synergies of brain activity stimulated by recurrent word use.

How do we hold essays together? How do we hold anything together? As, we, for example, scroll through Facebook, horrified at the recursive images of the latest school shooting, we search for answers. Facebook makes leitmotifs of the world, showing the same twenty video images posted by a thousand different friends over and over again. We see nothing but videos of this horrible tragedy, taking place in this horrible country, plagued by horribly indifferent adults allowing their kids to be shot with horrible weapons. But then we read some statistics that show that even though school shootings are terrible, in terms of real deaths by guns, suicide is a much more pervasive tragedy. But you can't show pictures of suicides. When Anthony Bourdain kills himself a few months after the shootings at Marjory-Douglas, Bourdain quotations and still pictures of him fill the Facebook pages all over again—a different leitmotif, a different reaction. The collection of videos produces action from the Parkland students and calls to congresspeople. The Bourdain suicide led to personal reflection—why did we love him so? Why did someone so successful feel like death was the only solution? We post our phone numbers and invite the potentially suicidal to call.

What I'm looking for is not just one hit that creates a cascade of associations. Like social media, I want those associations to make a pattern. I can't just expect one cue ball to do all the work. I can't expect my reader's brain's billiard balls to be lined up just right. And I am a crappy pool player. I'm thinking something more scattershot might make a better metaphor. But I'm not into pheasant hunting either. What am I good at?

My aunts were good at boyfriends. I wanted to be good at boyfriends. My aunts took me and my sisters to Wild Wave Waterpark and Lagoon. They were the young aunts and the fun ones but for some reason, they churned through a lot of guys. Because my parents were still married and we were from Utah and though not LDS anymore, the prevailing wisdom about marriage was that marriage was a good thing. My aunt Brooke, although married often, became my hero. She was the first one in our family to get her PhD. The first one to divorce her third husband, after he bailed on her, and go it alone for a while. She moved to Denver, which is like Detroit in that it starts with the letter D and there are bricks there and elementary schools and factories. At some point though, those images became part of her past, an essay that existed before she stopped looking for jerky boyfriends. Now, in Denver, she gathers different pools of images: Cherry Creek, 14th Street, The University of Colorado, at Boulder, train track, columbine. A new essay made out of new images that include no one named Randy and nothing brown or white at all.

6

The Brain Split in Half

Ira Sukrungruang

There is a gap. That gap is a mysterious space—the distance between planets, the microsecond of emptiness between the formation of a word and the voicing of it. In that space is nothing, which is a sort of something. In that space is expectant silence, though there is never silence. That gap is ever expansive. Is infinite. In it, a sun is shrouded in dark, awaiting discovery.

When discovered, the mouth of a universe will open.

It is easy to spin an abstraction. To create a universe, a planet, a life. Words possess that power. I can write a boy eats an orange, and you see a boy eating an orange. I can write my son eats an orange, juice dripping from his toddler fingers, and you see the juice of an orange dripping from a toddler's fingers. But there is a limit. Always a limit. You can't eat the orange I've written about. You can't eat my son's orange. Not because he won't allow it, though he is mightily possessive of his oranges. But because to eat this orange is to tear the page and shove it into your mouth and find out the orange is paper, dry and sharp paper. The enchantment, however, has happened. The brain has made the orange tangible. Has filled it with taste and memory of other oranges we've eaten in our lives. And sometimes the mention of the orange becomes something more. Becomes not the orange at all. But desire. Regret. Or, with my son, love. So you see? A universe does blossom like a mouth.

* * *

The brain, the organ, is ugly. I remember thinking this. I remember looking at brains in jars and thinking, *Ew*. Yet, it is responsible for everything. It makes meaning out of abstraction. But to think about the brain is to know virtually nothing. Dr. Lu Chen, neurobiologist, says, "We know very little about the brain." Nobel laureate Dr. Thomas Südof says the brain is harder to map than the human genome.

Still, it is difficult to consider something so ugly as the thing that necessitates our lives. It can be hard to see beauty in biology.

We are personae. We allow people to see whom we want them to see. We build our exterior—the way we walk and hold ourselves together. We formulate a language that is shaped by the context of the environment. If I am in the south of Chicago, I talk the south of Chicago. I act the south of Chicago. The south of Chicago is in my slumped posture, is in the pockets of the jeans I dig my hands deeply in.

Once in a bowling alley with neighborhood friends, a kid our age—about seventeen—bowled next to us. He had a ponytail and wore a tied dyed t-shirt. He laughed and made conversation with me. I laughed and made conversation with him. His laugh must have been sparked by the best pot of all time. I laughed, too, with that same high. When he spoke, his words were tinged in the languid lyricism of a stoner. So were mine.

After he left the alley, my best friend said, "It's like you're two people. The one with us and the one with him."

* * *

There is something valuable about the severing of the self in writing. You control which self gets on the page. Which self carries a story. Which self is needed. For example, in one essay, the self will smoke a cigarette, because he smoked a cigarette all the time and this story is about bad boys and bad boys—including this particular bad boy—smoke cigarettes. In another essay, the self cradles a wounded kitten, found in an abandoned desk outside the Thai temple in Chicago, because this self is wrestling with the idea of tenderness and loss and violence. In this essay he does not smoke or say the word "fuck."

This is character formation.

I'm not talking about that.

I'm talking about the writer disconnecting from narrative impulse and, indeed, from the present moment, and instead sinking into the part of the brain that does not register time, that parses memory, sitting in a dark room and pondering impossibility. The part that relives, reimagines, and redefines. The part that remains a mystery, even to the self.

* * *

This is how I feel. It's all feelings. My math-minded friend would say, "You writers and all your feelings."

In the last year, I've lost two people, whose faces I keep seeing when my eyes are shut. Especially when my eyes are shut. They come to me when my eyes are open, too. But the visual cues of the world bring me back.

Sometimes.

Sometimes I stay with those faces. Sometimes those faces talk to me, but I can't hear them. Their lips move. They smile.

They laugh but emit no sound of laughter. I am left wondering what are they trying to communicate. Or whether they are simply telling me a story I can no longer hear.

* * *

I used to like story. I craved it. Wanted it. I needed to know what happened next. Now, my mind skips like a flat rock on a pond.

* * *

As I've gotten older, I don't think I believe in stories anymore. What I believe in is the silence after a story is told. That satisfaction. Like the sigh after a swig of soda.

* * *

When I was a kid one of my favorite movies was *The Man with Two Brains*, a 1983 comedy that stars Steve Martin, as Dr. Hfuhruhurr, a neurosurgeon. The movie is filled with slapstick, nonsensical moments, a reason why the child me loved the narrative. The child me waited for the next crazy situation, the next silly thing. The child me loved physical action and reaction, like the zany sobriety test police officers make Hfuhruhurr take.

Now, I watch the movie and think, here is a man in love with a brain and not the body. Here is a man who would murder a body for the love of a brain. And suddenly, I spiral into abstraction.

* * *

I know it's me that sees the brain as this ugly gelatinous mass. In the movie, Dr. Hfuhruhurr displays a child-like joy being surrounded by jars of brains. It is as if he is in a prairie of wildflowers.

My son is nearing two. His brain, this sponge, is ever in wonder. It absorbs language. It commands the body and all its recklessness. Everything is new. My son's brain, right now, is in the constant state of learning.

No.

Living.

Dr. Barnett, author of *Idiot Brain*, says, "The beautiful thing about a toddler's brain is everything exists in the gray matter." Uncertainty rules. It's not until children reach their teens that they begin to lose memory. And memory, this planet of fog, suddenly becomes a type of certainty, the thing that shapes our living.

* * *

My mother's brain, at 82, is falling quicker into the gaps. I tell her things and she forgets them. When I overhear her talking to her friend in Chicago, the conversation loops in on itself. My mother is old. Her friend is old. Their brains are old. *How old is your grandson? Wait, Sue died? You have a grandson? How's Sue? How old? Tell Sue I said hi.* It is hard for me to negotiate this change in my mother, who had been sharp and attentive and ever-present throughout my life. She still is. But now I repeat myself. My impatience creeps up on me. My impatience can be venomous and unforgiving. *I already told you, Mom. Remember? Why can't you keep this in your head? I've said this again and again. You are killing me.* My mother absorbs this.

Until she can't.

Until she says to me, *You won't have to deal with this for much longer. I will be gone soon.* Guilt then quiets my rage, forces my head down, forms a deep sigh that is an unsaid apology.

* * *

My brain has begun to stall, like my mother's. My brain, this control center of every action, is aging, is aged. It is spending more time within itself. Less time experiencing the textural world. It is like an old hermit. He comes out and the sun is so bright it chases him back in. My brain is in a constant state of losing. It's losing by the second. And in that loss, sometimes, a gap forms and I fall in. I will keep falling.

I like taking internet quizzes that tell me whether I'm a right-brained or left-brained individual. The questions put you in what-if situations. Like: Your boss gives you a shit ton of work to complete by next week. How do you go about accomplishing this task? (a) You jump from one task to the other until all is done. (b) You put them in a list of importance and complete each task one by one. (c) You organize them by importance and work on one until you get bored then move to the next.

There are no questions on the tests that ask why I'm taking the test. I do it in many ways to confirm my life decision to be an artist. To prove that my choice to live in a world of creativity instead of logic and certainty was the right one.

Because sometimes I doubt.

Because, honestly, I like math. I like numbers. I'm good at them. Always have been. You are looking at Harnew Elementary's Math Champ Master from second to fifth grade. I crave organization and structure. It might be the reason why the results of all the internet quizzes I've taken suggest that my left and right brain are in perfect balance.

* * *

The real reason I take these quizzes: I'm procrastinating from doing any work, especially writing about the spaces in the brain.

* * *

In sixth grade, when I was dethroned as Math Champ Master, I read book after book after book. I wanted to be with words more than numbers.

* * *

Quiz: When you open the refrigerator and forget what you wanted to get out, you ____________________:

(a) stand there in the cold chill, trying to recall what it was you were looking for.

(b) close the fridge and curse endlessly about how much your aging brain sucks.

(c) grab a pint of ice cream and say *fuck it.*

Now I think about the space that separates the halves. I think about the meaning of the story, rather than the story. The story is a closed fist. The meaning of the story is the moment the fist opens.

It is something I'm dealing with. I know not everyone feels the same.

A friend of mine, a fiction writer, says, "Stories give him sense in a world of senselessness." We follow a character as they evolve. By the end, time has made that character different. "Time matters," my friend says, "because time evolves us."

I don't disagree.

Another friend says, "She distrusts plot. She hates the word plot when talking about writing, refusing to teach plot." "I'm reading all these things about elves and dwarves. And they are doing elves and dwarves things. It's vital to the plot of the story, my students say. I say so what. I say I don't care. Plot is nothing.

Emotion is the true engine of story. But how do you teach emotion?"

A left brain equation: Plot + Emotion = Story

A right brain equation: Feelings × Feelings = Feelings^2

When I was a kid, I remember my biology teacher calling me a right brainer. "You are the artist type. When you dissect that frog, you don't see the parts. The organs. You see the story of that frog. You see a frog."

No, I saw the organs, too. I saw the anatomical makeup of that frog. My brain was in the process of severing the parts of the frog. I was a good student. The difference was—when I saw the heart, I saw it beating. When I saw the lungs, I saw them breathing. When I saw the testes attached to the kidneys, I laughed because I was fourteen and the word testes made me laugh.

But I won't lie. My brain saw the frog come to life, gather into itself all we had taken out. Sewed his holes up with a magical needle and thread and bounced away to a pond where other dissected frogs go.

* * *

The people I've lost and loved exist in story. They do. But a story with holes. A story that is more sentiment than image.

* * *

My son loves storybooks. I have read the same story to him for the past three months. The story is about a gorilla saying goodnight. My son knows what will happen next, knows how the gorilla will open all the cages of the zoo animals so they can escape and sleep in the zookeeper's house. Yet he laughs as if he is viewing it for the first time. I, on the other hand, am tired of this book. I do not tire, however, of my son's feelings for this book. I have put his joy into a box in my brain. Labeled it *all the things I do not want to forget*, fearing one day it will evaporate, fearing one day the universe gone dark.

Biology lesson: That gap between the hemispheres of the brain has a name: Subarachnoid space. In it are spongy tissue and cerebrospinal fluid that cushions the brain and carries waste away. I want to make a big deal out of this space. I want to say that space is where the magic happens in writing, where what is lost goes to be found. But everything I've read, every article I've looked into, says nearly the same thing. It's simply a space: not entirely empty, though perhaps expectant. They also say, the space is a necessity.

7

When the Body Reads: Writing Sensory Perception for Reader Embodiment

Nancer Ballard

"There's a common fallacy that we have only five senses," notes scientist and craniosacral therapist Andrew Cook. In fact, we have somewhere between twenty-two and thirty distinct types of sensory perception including: sight, hearing, smell, taste, touch, balance, muscle strength, temperature, hunger, pain, fatigue, itch, satiation, the carbon dioxide/oxygen ratio (feeling of suffocation), pressure, gravity, thirst, circadian rhythms, the gag reflex, proprioception (relative body and limb position sensed by muscles and joints), kinesthesia (sense of motion), direction, acceleration and deacceleration, force, and vibration. Some scientists also include a sense of effort (the perceived force needed to lift, move, or manipulate an object), a sense of agency (the sense that tells us that when we lift a tea cup it is our own arm that is doing the lifting), rhythm/oscillation, electromagnetic vibration, and the passage of time. In a less-scientific way we also use the words "sense of . . ." to describe perception-emotion-behavior routines, as in having a "sense of humor" or a "sense of obligation." Some of our senses, such as smell and sight, respond to external stimuli; others, like muscle strength and thirst, are generated internally. Still others, such as balance, vibration, and temperature, can be initiated by internal or external stimuli.

Our sensory perceptions often involve multiple parts of the body working in concert. We "hear" through our skulls as well as our ears. The semicircular canals needed for balance are located in our inner ears, but our ears get help from our eyes, and we make constant micro-adjustments in muscle tension throughout our bodies. Many people can distinguish a dozen distinct types of pain including throb, ache, pierce, crush, grind, and "pins and needles." Others are sensitive to electromagnetic radiation, the "peri-space" envelope that surrounds our body proper, or subtle vibrations that allow them to "see" behind their backs. Humans also use their individual senses in diverse ways. For example, our eyes recognize friends, locate objects we intend to manipulate, orient us in time and space, and detect fleeting movement that could signal danger and automatically activates a fight or flight response.

Emotions are also woven into the tapestry of sensory perception. Fear, embarrassment, excitement, and all our other feelings are built on multiple sensory cues processed in different parts of the brain. When we're frightened our hearts beat faster, our respiration speeds up, our limbs tingle, and our mouths go dry. Physical sensation tells us that something important is about to happen, and our bodies produce more sensory-rich perceptions as we start to respond.

Creative nonfiction writers use sensory perception to affect a reader's experience in multiple ways. A reader can be induced to vicariously simulate the action and sensations of a character. Reading can activate the visual representation areas of the brain so that a reader can imaginatively "see" a place, a character, or an object as if she were observing them from outside the story. A reader's brain and auditory apparatus can be activated to give a reader the impression she is hearing a character or narrator speak. Or, a reader can activate her own "inner voice" and "inner ear" to help her make sense of a passage. In each case, the reader is spurred to physically participate in stories spun from words on paper. How does this magic happen?

Many of us naturally assume that that our senses' job is to enable us to accurately perceive reality. However, sensory perception is much more narrowly focused. According to Dennis R. Proffitt, a psychologist at the University of Virginia, and Donald D. Hoffman, a cognitive scientist at the University of California Irvine, what we see, hear, and feel are largely determined by automatic moment-to-moment assessments of the actions that our surroundings are prompting us to undertake. For example, when I want to lend a friend my copy of Ellen Meloy's *The Anthropology of Turquoise*, I scan my book case for a paperback with a watery blue-green spine and Meloy's name. I'm oblivious to the forest green box of printer ink cartridges that sits on an adjoining shelf; it might as well not be there. In other words, we don't perceive the world and then decide how to respond, we perceive the world based on what we might need to do (or avoid doing) next.

When we initiate action, neurons in the premotor area of our brains' parietal lobes activate the areas of the brain needed to carry out the action sequence. I spot the paperback book I'm looking for above my head. Pulling the book down requires me to drag the step stool over, open it, and climb up. I've done this many times, so my brain begins the physical mind-body routine needed to set up the stool without conscious instruction. Even when we watch a friend climb on a step stool, play tennis, or eat an ice cream cone, our bodies activate or "mirror" the neurological motor routines that go with that action.

Our ability to internally mirror another's actions is very useful, maybe even necessary, when acquiring new skills, such as how to operate a smart-phone or drive a car. Scientists have discovered that *reading* action-related words also prompts our bodies to physically mimic the actions we are reading about. When you read the word "grasp" (so long as it is being used literally, as in grasp the step stool rather than grasp an idea) the areas of your brain responsible for initiating hand movements are activated, as if you were watching someone grasp a step stool, or even, were grasping it yourself.

When we read narratives the visual representation areas of our brains are also activated. Nicole Speer, Operations Director of the Intermountain Neuroimaging Consortium at the University of Colorado, and her colleagues have found that that when a character in a piece of literature picks up an object, both the reader's visual areas and the premotor areas are activated as if the reader were watching and feeling herself lift the object.

Silent reading also activates areas of the reader's brain associated with language production and language comprehension and the reader's throat and tongue muscles. Not only do we imagine and physically mirror the action in sentences such as "I stepped onto the stool and grasped the shelf to steady myself," we also physically respond to the words as if we were reading aloud or listening to the passage being read.

The kind and frequency of sensory cues that a writer provides have a significant impact on whether a reader will embody a character; picture an event while being psychologically situated outside the story-world; "hear" the voice of a character or narrator in her mind; or use her own internal voice and ear to "make sense" of a passage. If a writer wants a reader to embody a character, he must provide the reader with a sentient being (someone for the reader to embody), an environment (somewhere for that person to be), and a suggestion of sensation for the reader to simulate. Here is an example from Jon Krakauer's first-person account of a disastrous Mt. Everest expedition:

> Wanting to conserve whatever oxygen remained in the tank, I asked [Harris] to reach inside my backpack and turn off the valve of my regulator, which he did. For the next ten minutes I felt surprisingly good. My head cleared. I actually seemed less tired than I had with the gas turned on. Then, abruptly, I sensed I was suffocating. My vision dimmed and my head began to spin. I was on the brink of losing consciousness.
>
> Instead of turning my oxygen off, Harris, in his hypoxically impaired state, had mistakenly cranked the valve open to full flow, draining the tank.

In this passage from *Into Thin Air*, the narrator (Krakauer), the environment (the mountain), and the action (draining the oxygen tank) are bound together with urgent life or death stakes.

But prompting a reader to place herself in a story need not be as dramatic as having one's oxygen supply extinguished. Here is an example from George Orwell's *Such Such Were the Joys* in which the reader is invited to experience the confusion and shame of a young boy at boarding school:

> I was warned that I should be beaten next time, but I received the warning in a curiously roundabout way. One afternoon, as we were filing out from tea, Mrs. Simpson, the headmaster's wife, was sitting at the head of one of the tables chatting with a lady of whom I knew nothing, except that she was on an afternoon's visit to the school. She was an intimidating, masculine-looking person wearing a riding habit, or something I took to be a riding habit. I

> was just leaving the room when Mrs. Simpson called me back, as though to introduce me to the visitor.
>
> Mrs. Simpson was nicknamed Bingo, and I shall call her by that name for I seldom think of her by any other. . . . She was a stocky square-built woman with hard red cheeks, a flat top to her head, prominent brows, and deep set suspicious eyes. Although a great deal of the time she was full of false heartiness, jollying one along with mannish slang ("*Buck* up old chap!" And so forth), and even using one's Christian name, her eyes never lost their anxious, accusing look. It was difficult to look her in the eye without feeling guilty, even when one was not guilty of anything in particular.
>
> "Here is a little boy," said Bingo, indicating me to the strange lady, "who wets his bed every night. Do you know what I'm going to do if you wet your bed again?" she added, turning to me. "I am going to get the Sixth Form to beat you."
>
> The strange lady put on an air of being inexpressibly shocked and exclaimed, "I should think so!"

In this excerpt from Orwell's essay about the culture of English boarding schools, Orwell combines environmental cues (filing out of a dining hall), the young boy's external visual sensations (the appearance of the headmaster's wife and her visitor), internal sensations ("difficult to look her in the eye," feeling guilty, and the anticipation of being beaten), and action (leaving the room, standing, the headmaster's wife turning, and the visitor's expression of shock). Although the boy's physical action is relatively simple, Orwell's use of internal and external sensory cues helps place the reader in the shoes of a child being shamed by the wife of the headmaster in front of an unknown visitor.

In *The Anthropology of Turquoise*, essayist Ellen Meloy leads the reader to embody her scene with almost no character action.

> On the roof the sun pours down in a white flood. The black felt absorbs the heat just short of a meltdown. Along the river bottom beyond the roof, the coyotes sing with great fervor, perhaps a great opera celebrating the delicious flavors of grasshoppers, rabbits, and mice.
>
> When age fixes me to a chair on the porch, as I am fixed to this roof, I hope I will have the courage to wipe away the drool and pay attention. Shuffle my bent back and hummingbird bones down to the river to see what is flowing there and weep over the memory of how oars felt in my fists when I ran a raft through a high-water rapid, how a person can carry the river's muscle in her own. Adapt an experiential shrewdness that tricks everyone into believing it is wisdom. Crack some land vandal over the head with the collected works of Henry David Thoreau.
>
> When you can no longer power your way through it all, when an active life slows down to prayerful attentiveness, is there still room for both the fever and the mirage?

Meloy transports the reader into her scene and then into the scene-within-a-scene by situating the narrator (and the reader) in her present surroundings on a roof in the Southwest and then using the same verb ("fixed") to situate the reader and her future self in a porch chair from which Meloy animates an imagined future with verb-rich reflection.

A sentient being that the reader can embody and some literal, remembered, or hypothetical action may be necessary for a reader to inhabit a scene, but it isn't sufficient. To engage a reader's sensory-motor apparatus a writer must introduce, and continue to introduce, sensory cues. In the sentence—"Anneli set the table."—there's a character and an action but nothing to suggest a sensory experience, so it's virtually impossible for a reader to place himself within the story. Notice the experiential difference between reading that sentence and reading the actual passage from M. F. K. Fisher's *As They Were* in which Anneli appears:

> Anneli [the cook's assistant] set the table, and then came back with a net and a platter. She swooped up a trout, held it by the tail, and before we could close our ears or even wince, had cracked its skull smartly on the sideboard.

While reading Fisher's description a reader is probably still observing Anneli, but Fisher's use of sensory cues has now invited the reader into the restaurant with the narrator and her companions as they shrink from sound and sight in response to the unexpected blow to the trout skull.

How much sensory cueing a reader needs depends on the circumstance. In everyday life our sensory perceptions are sorted and weeded below consciousness where our brains' perirhinal cortex neurons sort for novelty, usefulness, need-to-know, want-to-know, and don't-need-or-want-to-know. We consciously notice what we need to avoid and what we need to act—that is, danger and instruments of action. This is why I see the step stool when I need to pull down Meloy's book but not the box of printer ink, and why I see the box of printer ink but not the space where the book stood when my printer tells me I need to replace an ink cartridge to print a draft of this essay.

In life we often physically prepare for action *before* we perceive—as my need to retrieve a book from a high shelf prompted me to see the step stool. In narratives, a person or object is often introduced before the character interacts with it. In life we encounter thousands of objects and beings that are not relevant to our action intention and they pass through our minds unnoticed, mostly below consciousness. However, readers expect and assume that objects introduced in narratives are there for a reason—their existence in the story is already evidence that they are important for the character, the plot, or some other element of the narrative. If items were only introduced in stories after the character needed to use them, the story would feel contrived. In life, I may only consciously see the step stool when I need to use it, but I know it is there, whereas in a story-world objects and people don't exist until they are mentioned.

What can science tell us about the amount of detail needed to transport a reader into a story-world? Daniel Cassanto of the Max Plank Institute for Psycholinguistics and Anežka Kuzmičová of Stockholm University report that there are individual differences among readers (and even in a single a reader over multiple readings), but in general when a character interacts with an everyday object the reader doesn't need much sensory detail because he already has a mind-body repertoire for how to interact with it. When a writer provides a lot of detail about something used in a typical fashion, the scene feels tedious or slow and causes the reader to fall out of the story. When a character or the narrator is interacting with an unusual person or object (or an object is used in a very unusual way) more detail is needed so that the reader can imagine the person or the object's use. In *Pilgrim at Tinker Creek* Annie Dillard illustrates both points:

> I hear a roar, a high windy sound more like air than water, like the run-together whaps of a helicopter's propeller after the engine is off, a high million rushings. The air smells damp and acrid, like fuel oil, or insecticide. It's raining.
>
> I'm in no danger; my house is high. I hurry down the road to the bridge. Neighbors who have barely seen each other all winter are there, shaking their heads. Few have ever seen it before: the water

> is *over* the bridge. Even when I see the bridge now, which I do every day, I still can't believe it: the water was *over* the bridge, a foot or two over the bridge, which at normal times is eleven feet above the surface of the creek.
>
> Now the water is receding slightly; someone has produced empty metal drums, which we roll to the bridge and set up in a square to keep cars from trying to cross. It takes a bit of nerve even to stand on the bridge; the flood has ripped away a wedge of concrete that buttressed the bridge on the bank. Now one corner of the bridge stands apparently unsupported while water hurls in an arch just inches below.
>
> It's hard to take it all in, it's all so new. I look at the creek at my feet. It smashes under the bridge like a fist, but there is no end to its force; it hurtles down as far as I can see till it lurches round the bend, filling the valley, flattening, mashing, pushed wider and faster, till it fills my brain.

When Dillard first mentions the bridge she passes over every day, she doesn't describe what it looks like or what it's made of; these details are given only when the bridge has become unusual as a result of the flood. As she describes the flooding creek and the broken bridge, Dillard adds more internal and external detail (it takes nerve to stand on it; a corner is ripped away), and she saturates the passage with alliterative action verbs: recede, rolled, trying, ripped, buttressed, and hurtles.

Writers often imagine that a vivid image will sustain a reader for several pages, but scientists tell us this is rarely the case. The embodiment produced by action-object-sensation combinations is instantaneous but brief. Experience is made up of a multitude of discrete actions and effects. To keep the reader's body engaged a writer must regularly introduce new object-sensation interactions.

In *Autobiography of a Face,* Lucy Grealy offers a stream of sensory details that gradually reel the reader in close to her young narrator:

> Sunday afternoons in the hospital were the stillest and the longest—formless hours to be gotten through. With all the departments closed, there was none of the week's bustle. The familiar nurses were off, leaving us in the hands of unsympathetic aides who didn't care if we were entertained or not. In the stillness, the traffic on the street below sounded louder. There were more of the other patients' visitors to watch, obscure relatives who made the trip from out of town bearing useless flowers and ornately wrapped toys. But I grew tired of scrutinizing them, grew to recognize the swirling patterns and dynamics of every family that walked into the ward complaining of how hard it was to park around here, how long the elevators took. Some older brother or father would find a surgical mask and put it on and laugh, believing he was the first to discover this antic. I'd sit on my bed looking for words hidden in a jumble of letters or vainly attempting to put together an incomplete jigsaw puzzle I'd found in the game room. The stiff sheets made the bottoms of my feet red, and I was always in trouble for not wearing my slippers when out of bed. A tart smell drifted down the hall from the sluice room, where they cleaned the bedpans and kept the sterilizer.

To capture the girl's boredom and detachment, Grealy constructs her description of the hospital with sight and sound, our most remote senses. Toward the end of the passage she brings readers into the room by describing the girl's chafed feet and evoking the sharp repellent smell of a used bedpan. In *Pilgrim at Tinker Creek*, Dillard continues to engage the reader's body by shuttling between human and nonhuman action, fusing the two by comparing the flood waters to a body part (a fist). This is followed by a string of action verbs describing the water (which the narrator and the reader can only observe), and then Dillard invites the reader back into the narrator's body as the lurching, mashing creek "fills my brain."

Although reader embodiment requires a writer to rouse the reader's internally oriented senses and premotor responses, references to such sensations tend to be less numerous and less explicit than externally oriented sensory cues. This tracks with our experience of the world because so much of our internal sensing goes on below consciousness. We periodically become aware of a muscle twinge or change in respiration, and can describe them if asked to do so, but most of the time we are more focused on the person or object with whom we are interacting or going to interact. We focus on our internal sensations when our actions are thwarted, when we are experiencing an unusual condition such as dizziness, pain, or unexpected joy, or when we are attempting to prevent our bodies from reacting.

Thwarted action, unusual conditions, and trying to resist visibly reacting are, of course, an integral part of most of our stories. Internal sensations such as dizziness, pain, and muscle strength place the reader within the body of the character experiencing the sensation unless it is obvious from the text that another character is observing.

Touch is our most intimate and most fear-sensitive externally oriented sensation, for it requires contact. Touch may also be our most miraculous sense—from an evolutionary standpoint everything began as touch; it's the one sensation that all live beings possess; and our sense organ for touch, our skin, covers the entire body. Touch is simultaneously an internal and an external sensation. When we touch another person, object, or ourselves, we are also being touched. Indeed, we usually describe touch as a recipient rather than as an actor. A snake skin feels dry and fragile; a baby's breath feels warm and moist. References to touch tend to place the reader in the recipient's position unless the writer makes explicit that touch is being observed.

Smell is our next most immediate external sensation. You have to be close to smell things (except forest fires and paper mills). Smell references usually propel the reader into the perceiver's body. Touch and smell are wired directly into the lower "emotional" areas of the brain where our neurological circuits are faster than reason and prompt an immediate response in the reader. Curiously, although we can distinguish a trillion different odors, we don't have many words to describe smell. Instead we identify most smells by their source. Readers can conjure the smell of insecticide or a sizzling fish even if they can't precisely describe their odors because emotion and nonverbal memory play such a large role in recognizing smells.

For obvious reasons taste also places a reader inside the imagined perceiver. Taste is a combination of smell, chemical reactions in the taste buds, and texture, any of which can be highlighted to produce a slightly different effect in the reader. Because we are physically averse to eating things that could poison us, and we also associate food with comfort and safety, our physical and emotional reactions to imagined tastes dwarf pure taste sensation.

Hearing is essentially vibration, a variation on touch. Although we can hear things that are too far away to touch, taste, or smell, hearing and touch are closely connected. Sound vibrations can be felt in the skull, the solar plexus, and the soles of our feet, and we describe music as being able to "touch" the soul. Hearing is particularly intriguing because it's associated with secrets, which can be simultaneously intimate and anonymous or draw attention to the ephemeral territory between the known and the unknown. Unless hearing is paired solely with sight (our most remote sense), it usually places the reader in the perceiver's body.

Vision is our most complicated external sense. We can see from the perspective of a participant or from afar as a spectator. We see objects at a distance we can't otherwise sense and can also examine people and objects at such close range we can't perceive their whole. We use vision to orient ourselves in time and space. When writers describe a person or object from a perspective that's closer than a person would typically see them in real life, the reader often feels resistance, distortion, or danger. Our peripheral vision is sensitive to movement and can make us instantly respond to potential danger or reward. Our eyes begin life as part of our brains and remain associated with mental imagery and understanding throughout our lives. "I see what you mean," means we understand what someone is saying. We also use vision to maintain a continuous sense of ourselves and the world whereas most other individual sense perceptions are short-lived. This is why a reader will read (and physically respond to) a scene containing multiple types of sensory cues and recall the scene first as a visual image. Even the word "imagine" is based on the word "image" and bows toward the visual.

One might think that with more than twenty different senses, it would be easy to keep a reader embedded in a narrative, but more description doesn't necessarily yield more vivid embodiment. In fact, extended description can cause readers to fall out of the story. Consider Alfred Russell Wallace's description of the parrot:

> As a rule parrots may be termed green birds, the majority of the species having this colour as the basis of their plumage relieved by caps, gorgets, bands, and wing-spots of other and brighter hues. Yet this general green tint sometimes changes into light or deep blue, as in some macaws; into pure yellow or rich orange, as in some of the American macaw-parrots (Conurus); into purple grey, or dove-colour, as in some American, African, and Indian species; into the purest crimson, as in some of the lories; into rosy-white and pure white, as in the cockatoos; and into a deep purple, ashy or black, as in several Papuan, Australian, and Mascarene species.

For many readers, Wallace's description of parrots is likely to swamp their ability to assimilate detail while maintaining a mental image of the whole. Extended descriptions, such as the excerpt from Wallace's *Tropical Nature and Other Essays*, frequently rely heavily on visual details, which tend to place a reader in the position of an observer rather than a participant. Without other nonvisual sensory cues that place the reader in a scene, the reader becomes a detached observer looking at a static landscape, object, or parrot.

Also, even multisensory descriptions can feel artificial or forced if they don't track the way we experience the described person or object in life. In life we usually don't see a 360-degree view of an object or know where it came from; we notice people and objects because we need to interact with them. When the reader is asked to imagine a person, place, or event in more detail than seems pertinent to the progress of the story, he becomes impatient. Details that are not linked to action are also quickly forgotten. The reading body leans toward action, just as in life our bodies incline toward purposeful activity.

It can be disappointing for a writer to realize that he finds the examination of a peculiar or beloved object more engrossing than his readers do, but there are several reasons for this. First, the writer often has the artifact at hand, so the writer has the actual experience of touching, holding, or smelling the item, even though the narrative doesn't require a character to interact with the object with such attention. Second, the writer has a purpose for interacting with the object—namely to write a description of the object whose purpose is to evoke a vivid experience within the reader. But the reader doesn't have this purpose; the reader's interest in the object is circumscribed by his own previous personal interactions with similar objects and the character's interaction with it in the story.

Another reason that description tends to detach the reader from the story-world is that descriptive image-building relies on the reader's frontal lobe rather than the faster premotor/emotion circuits that are activated by a character's need to act. The reader's pace slows as the image is being consciously assembled or elaborated in the reader's mind. The reader may even become aware of having to invest cognitive effort in the conjuring process. When the forward progression of a story pauses for reflection or description, the description can feel like a semiautonomous chunk of text rather than an integral part of the story. (This is why readers often feel like they can skip description without missing much.) The quickest way to bring a descriptive passage at least partially into the realm of inhabited story is to include some sensation, such as smell, taste, or an internal sense, which must be experienced from inside the body. In *Autobiography of a Face*, Lucy Grealy brings the reader close by evoking the smell of the bedpans in the sluice room down the hall. In M. F. K. Fisher's passage from *As They Were*, the reader is brought into the scene when Fisher's narrator describes the action she and her dining companions would have taken—wincing and covering their ears—had they known the young cook's assistant was going to crack a trout's skull in front of them. Ellen Meloy brings the reader into the story with the word "delicious," an adjective that implies a subjective internal experience, as she describes the coyote's imagined songs about grasshoppers and mice. In each of these examples, the reader's detachment falls away when sensation is used to locate the reader within a physical body.

Descriptions of unusual people or objects can pique a reader's interest apart from the story. As humans, we are drawn to novelty, especially if the unusual version defies our ordinary understandings and makes us rethink our mental templates or shows us something intriguing about a character. Here is an example from *Lincoln's*

Melancholy, in which Joshua Wolf Shenk describes his attendance at an annual convention of Lincoln actors.

> Four years ago, on a lark, I went to spend a weekend with the association of Lincoln Presenters, who were having their annual convention in Beckley, West Virginia. Of the several hundred members, forty-four Lincolns showed up that year—men dressed in black suits with stovepipe hats and beards shaved about the chin. Some of them were short and thin, some were tall and hulking. Some had real beards with gray hair, others had false beards made of black hair. One of the Lincolns was in a wheelchair, with an oxygen tank. One looked like Elvis Presley dressed up as Lincoln—sideburns, sunglasses, everything but the gold lamé.
>
> I went to Beckley because I thought I could learn something from these men that wasn't in books. . .

In this passage Shenk uses description of the unconventionally attired Lincolns to challenge the narrator and the reader to examine their stereotypical views of what Lincoln should look like and to consider whether the cut of a beard or a particular height is essential to understanding a president that suffered from depression, and also emancipated slaves, penned the Gettysburg Address, and governed a country through civil war.

Writers also draw upon readers' natural sensitivity to novelty when they use figurative language. Metaphors use a familiar image to highlight some aspect of a more complex or less familiar object. In the description of the flood, Annie Dillard compares the river to a fist. Ellen Meloy compares the howling of coyotes to an opera and her brittle bones in old age to hummingbird bones. The unexpected but familiar metaphoric images leap into the reader's mind and then bind to and amplify the principal, more-difficult-to-imagine image—

an arching creek, sentient coyotes, and the multidimensional frailties of growing old.

When a reader feels there is purpose to the description apart from the progression of the story, the reader can momentarily step out of the story and find unexpected fascination or amusement in a semiautonomous description. If the description offers the reader independent marvel, pleasure, insight, or humor, the pause can delight the reader's mind and senses and may be remembered long after the book is closed.

Until the nineteenth century, the embodiment of a human voice through verbal expression and oral style was more important than visual detail because people listened to sermons, public lectures, and stories more often than they engaged in silent reading. Also, until relatively recently narrative poems and stories tended to involve the mythic, large groups of people such as armies or tribes, and the sublime, rather than individuals' everyday activities for which listeners had automatic preparatory mind/body routines. Although most modern readers describe feeling embodied in a story in visual terms, they also find dialogue to be a vivid part of the in-scene experience. My unscientific survey of fellow nonfiction readers also leads me to believe that the "imagined relationship" (rapport) with a distinctive narrator can be as pleasurable and evocative as immersion in a plot.

Auditory or speech imagery is a silent reader's imaginative impression of "hearing" words as if spoken by a character, the narrator, or the reader himself. In "embodied speech imagery," the reader embodies a character or narrator's voice so that the words mentally "sound" as if they are being spoken by another. Embodied speech imagery is often cued by oral style—a distinctive accent, diction, attitude, or orientation toward the world. Listen for the voice of narrator Jane Brox describing visits with her aunt in *Clearing Land*:

> Always at the end of my brief visits, after I'd filled the air with patter about small things—the growing season, the weather—and had repeated myself to keep the air stirring, after I'd said I should be going, she'd query me: "What's your hurry?" I could feel the silence I would leave behind me.

Although the dominant feature of sensory speech imagery is "hearing" another's voice, in Brox's passage I can also feel the consonants behind my teeth while "hearing" the voice of the narrator. Her aunt's understated question "What's your hurry?" echoes in the hollow of my mouth. Embodied speech imagery, like embodied visual imagery, occurs spontaneously and without reader effort.

Embodied dialogue tends to be composed of short sentences and relatively simple grammar. Character and narrator reflections are often more complex, but the longer and more complicated the sentence construction and concepts, the more likely it is that a reader will have to use his or her own real and imagined vocal apparatus to parse the meaning of a passage, and the distinctive voice of the character/narrator will fade into background. As a story progresses, readers often become accustomed to the voice, diction, and attitude of a character or narrator so that more complex diction and syntax can be processed by the reader as embodied. An embodied voice may sound stronger or clearer upon a second or third reading after the reader is familiar with the content and diction.

There is another kind of auditory imagery in which the reader "hears" the sound of her own voice. Scientists refer to this as "subvocalized auditory imagery" because the reader isn't speaking but employs her own "inner voice" as well as her "inner ear." Subvocal auditory imagery may be a throwback to our early elementary school reading days or a historical time when people were just learning to read silently and "used their lips" to help create sense units out of words and translate them into meaning. Readers often switch into subvocalizing when they meet large chunks of exposition or reflection, or long sentences with complex syntax or complicated concepts. Although the reader isn't talking or hearing words spoken aloud when engaged in subvocal auditory imagery, the areas of the brain involved in voice production and in auditory reception are activated. Subvocal auditory imagery also activates the reader's vocal cords and mouth and larynx muscles. At a physiological level, the reader is "speaking" to herself as well as "listening" as she endeavors to make sense of an important or complicated idea. Here is an excerpt from Rebecca Skloot's *The Immortal Life of Henrietta Lacks* in which readers are likely to shift into mentally reading aloud to themselves as they grapple with the concept of cell immortality.

> After years of disbelief and argument from others scientists, Hayflick's paper on cell limits became one of the most widely cited in his field. It was an epiphany: scientists had been trying for decades to grow immortal cell lines using normal cells instead of malignant ones, but it had never worked. They thought their technique was the problem, when in fact it was simply that the lifespan of normal cells was preprogrammed. Only cells that had been transformed by a virus or a genetic mutation had the potential to become immortal.

Subvocal auditory imagery (mentally reading aloud to one's self) draws upon a reader's physical and sensory vocal apparatus to bolster reading comprehension. The physicality of the effort is apparent in the etymology of the word "comprehend," which comes from the Latin words for "com," meaning "completely," and "prehendere," meaning "grasp" as in grasp with the hand (prehendere is also the origin of the word "prehensile").

To the reader, subvocal auditory imagery is similar to thinking to one's self, except that there is often more of a sense of effort because the reader is working through someone else's thoughts. In reflection, the narrator or character may be struggling to make sense of something herself, and the reader effectively lends her voice and effort to assist in the endeavor. Does that say what I think it says? What exactly is going on? What does the character mean by that? The reader uses her own articulatory apparatus to trace a rhythm or parse a phrase. She may even break into a whisper or sense that her lips or vocal cords are vibrating (which they are). At literary readings, I sometimes see members of the audience sitting with a finger or hand touching their throat or lips, so that they can "hear better." Like readers, they subconsciously (or consciously) know that they can hear more clearly when they feel physically connected to the words.

Imagery that leads a reader to inhabit a scene or a character often steals the limelight in writers' discussions of sensory perception, but each of the ways in which readers process sensation has its own benefits. We are drawn to in-scene immersion because inhabiting a world and characters with something important at stake makes us feel alive without the actual risk of being in their position. Immersion in other worlds and people's perspectives allows us to experience firsthand their dreams and losses, thrills and struggles. We can also embody cultures and time periods that are impossible for us to directly experience in our own lives.

Through description readers can examine foreign territory or unfamiliar aspects of familiar terrain. Description can provide humor or a new way of looking at someone or something that challenges our stereotypes and assumptions. The pauses in embodied immersion provided by description and character reflection give readers a chance to think about the implications of a character's actions or events in the world beyond the story. Description can change our perspective by bringing us in close or encouraging us to view a subject in a larger historical context. Description also cues a reader that something important involving this place or object is going to happen because it is worthy of notice in a particular way.

For longer than we have been silent readers we have listened to spoken voices with our bodies. A distinctive narrator gives us a sense of companionship and can help us through difficult physical, emotional, or conceptual places. Our ability to embody characters' and narrators' voices is like having a magic telephone through which we can hear other people's confidences and thoughts. It gives us an outlet for our own unexpressed dreams and feelings and supports impulse control. Embodying others' voices encourages empathy and helps us understand others' theories of mind and decision-making.

Subvocal auditory imagery also has an important place in creative nonfiction. When the reading gets rough we lend our voices to help ourselves and our text companions along. We travel at a more measured pace which helps us comprehend complex ideas, abstract thought, and extended metaphors. And when a writer gives us new insight we feel the gift in our bodies—the bloom of thinking someone else's brilliant thoughts as closely as you think your own.

8

LENS: A Lyric Meditation

Katharine Coles

1.

The eye, fragile and durable, tells time, which is, in essence, a function of the brain, figuring and reconfiguring our reality. Consider the miracle that any of us can see at all, much less distinguish 10 million-odd colors—but the eye is so useful it evolved independently, scores of times, in species that might as well be alien to each other, except we all evolved on a globe spinning in the light of a sun. In the human eye, the iris regulates the amount of light that can come in via the pupil, the aperture that tightens in response to brightness or aversion, eases in the presence of darkness or hallucinogens or love. My driver's license says my eyes are blue, by which it means my irises, but they can appear gray or green or any combination of the above depending on what I am wearing, the quality of light reflecting off of them, my mood. From photos and memory, I know I have my grandmother's eyes, which means not that I *see* like her but that I *look* like her.

Behind the pupil in its expansion or contraction, we find suspended the *lens crystallin*, which refracts light to be focused on the retina, which in turn creates the image the optic nerve shunts back into the brain and a cluster of cells so tiny we had to invent the scanning electron microscope to see it. When I was in high school, we could name the mind's eye, but only our intuition told us it existed.

We experience our vision in a way that makes it seem like a direct and accurate apprehension of the world. *Seeing is believing*, we say. But any face I believe I see before me is only my brain's reconstruction of light shuttling through the vision system's assembly line. Thus, science, which relies on proof, gives us so many reasons to doubt ourselves.

The lens, by the way, is metabolically active, sugar-loving though not voracious, not nearly as demanding energy-wise as the heart, say, or the brain. We also call the lens the *aquula*, which Wikipedia, another kind of window I open onto what may or may not be real, reminds me is Latin for "little stream."[1] The lens is bathed and moistened by the aqueous humor in front of it and the vitreous body behind. That the eye is a tiny sea, awash, has nothing to do with our propensity to weep. Tears come from elsewhere. I have thought of the lens as stable and ongoing, but, as I should know from experience, it grows and changes shape. With age, it can develop cataracts and grow progressively opaque, especially when the eye is exposed to too much bright light. Getting older, we need more and more help to see.

The lens is nerveless and bloodless. It lacks connective tissue. Depending on how you slice it, its layered laminae formed from crystallins, it can look like an onion or a honeycomb—but we don't want, do we, to rerun that particular film in the mind's eye, the razor crossing. While the rest of the eye is related materially to the brain, the lens is cousin to skin and sinus, fingernail and tooth.

Depending on what kind of animal you are and what you need to see, and whether you look for it through air or water or mist and in what kind of light, the lens may be elliptical or spherical. Pads or fibers may focus it, or muscles move it forward and backward or tilt it in the eye, an idea that makes me dizzy. A fly's eye has many lenses, as may a mollusk's.

The mollusk has its own story.

2.

I digress. And digress again. The world is too much with me.

3.

I am considering having my own lenses replaced with new ones, human-designed and made to be as good as the old. Better, if I get bifocals. Of course, at my age my original lenses tatter and grow shabby, but I am thinking about replacing them before I absolutely need to, according to my insurance company, while I can still read and drive at night.

I've written the story of my eyes before, in a poem. By the age of seven, I was so myopic I saw nothing without an obscuring halo; I couldn't even see the clock, its hands glowing, at my bedside. My teacher, a kind woman, told my less-observant mother, "I don't think she can see the blackboard." Many years later, a radio producer dug up a photo that had been in the newspaper not long after: my white blond hair blowing from under a patterned silk scarf tied under my chin, blue cat-eyed glasses, "a young Jackie Kennedy," the host observed to me.[2] By the time he knew me, I was a poet, known for seeing. Back in 1968, I was marching with my parents to honor Martin Luther King after his assassination. I had insisted they bring me, though I was on crutches for a broken leg. I hated those glasses, which distorted the world by prescription, so I could see it clearly.

Occasionally, the difference between what I could see through my glasses and the blur in my peripheral vision confused my brain and sent me tumbling. By my late thirties, my nearsightedness and astigmatism could barely be corrected with any lenses. So I let a surgeon pass light across my eyes, "So I too will see by the numbers,/ Cornea peeling, slice by measured slice."[3] The night after she had done the first eye, I woke in the small-hour darkness. "A luminous thing," I would write, "time's face/presents itself." For the first time in more than thirty years, I could see the clock unaided.

But for weeks when I woke in the night, I continued to make my way to the bathroom with my eyes closed, "feeling my way by hand/ through the familiar dark"—or with my eyes open but strangely unseeing. I had to "remind my brain to use them."[4] "Remind," a word I used in the poem as we don't usually use it, in the material sense. There was still a neural pathway there, so seldom traveled it petered out in the weeds. *Wake, see.* I had to work to reopen it.

4.

What is the difference, telling this story through the medium of prose as opposed to poetry? As a lens, what does each genre permit me to say, so to see? Prose, when it does not self-consciously deploy lyric distortion, can offer an illusion of clarity, as if we are experiencing events directly, unmediated.

Poetry, even when it arrives in sentences rather than lines, provides tools for compressing, for bending, for linguistic distortion, I hope to let me see clearly, like a corrective lens. Poetry does this through figures like metaphor, which might seem to deceive, but which in this case brought me through the invisible figure that occupies "remind" to its literal origin—a use that, because poetry is duplicitous, because it is the bi- or tri-focals of genre, sits alongside my unthinking deployment of the word. In bringing something into memory, I return it to my mind—and am myself returned to it. I remember that my mind is not separate from my body, however I feel it is; it is my body, all and only. At heart, *Remembering* is material.

Here, in prose, I tease these relationships out, abstract them to make them visible and clear, hoping you won't miss any of the pleasures or annoyances I've presented. Poetry allows me to put certain kinds of pressure on the language without explanation, to count on the reader to ferret out densities. Even here, writing in prose, I want to reach for that language, and often do. But prose provides me with the space to elaborate, and the space to digress. Here, I make room for my beloved teacher, Mrs. Brothers, to appear, and now her daughter, my friend Ravonne, who, through the lens of memory, remains gawky, though perhaps she was not. I think I was not an easy child for a teacher to love—like now moody, interior, deeply day-dreamy, given neither to finishing my homework, nor to performing it as instructed. "Even now, too often,/I close my eyes to see."[5]

Yet Mrs. Brothers saw me, and made the right diagnosis, perhaps because she saw a little of Ravonne in me. Decades later, radio producer Elaine Clark digs up my history, and Doug Fabrizio, the host, transforms an eight-year-old made self-conscious by her glasses into a figure of unconscious glamour, a foreshadow and vision of the poet she would eventually be seen as.

After my first eye surgery, I took to calling myself a miracle of science. Having always been a distance swimmer obliged to stick to the pool, where I could follow the big black line from end to end, counting strokes so I wouldn't hit the blur of wall, I took up open-water swimming. I stood for hours at a time at my window, looking up the canyon outside, noticing distance and detail, the clarity of the hawk hovering in an updraft focusing its own hawk-eye on the smallest movement on the ground. Even now, I am fully aware of my looking, full of it; "I am living," as I say in yet another sight-obsessed poem, "on the skin of my eyes."[6]

5.

Are all my poems about seeing? Perhaps lap swimming and its endless returning made me a poet; maybe being able to swim in open water led me, around that time, to plunge into prose. I can't say.

When I went to Antarctica, I was given lenses to protect my eyes from too much sun. But I wanted to experience that vast space, its reflections, refractions, and mirages, all the deceptions light plays, unmediated. Not wanting those lenses between the view and me, I kept taking them off, until the sun and wind burnt my eyes.

Prose lets me tell you outright, as my vision begins to dim and blur again, that I am saving my money; I am willing to pay out of my own pocket to arrest this process.

6.

In perception lies the crux of story, the crux of lyric. By perception I mean more than literal vision, but vision is a good enough place to start, a place to lose yourself, in an organ as small and complex as the eye, and a place to finish, even as we distort our vision to extend it, so we can see what grandeur lies beyond us, too small, too distant. Yes, our telescopes now as likely as not act like ears, great arrays of them turned together toward specific zones in the sky, listening; but when I hear the word "telescope" I still think of Galileo, his mind through his eyes cast upward by glass he ground himself, "levitating into the night sky," as Rich wrote of Caroline Hershel, "riding the polished lenses."[7] I have seen, in Florence, his instruments, whose lovely wooden shapes bear his touch and so bring him forth physically, if not literally, into our time.

7.

Here, now. The scientific, the personal, the literal, the metaphoric. The bodies we have and those we think we have, sailing our minds through our pasts, our visions, our words, our projections.

Literary language is full of the I, of the eye. The poem stacks up images. The story unfolds. Whatever we imagine, present or not, the visual centers of our brains catch fire.

I can feel myself becoming pedantic, saying again to my poor graduate students, trying to get them to keep more than one idea in mind at once, "your mind isn't mystical; it's meat."

So are my eyes. And my images of iceberg, redtail hawk, next month's eclipse, all of you: remember where I keep them, in my mind's eye, that cluster of cells so small you can scarcely imagine it, with me.

9

The Heart and the Eye: How Description Can Access Emotion

J.T. Bushnell

I once burst into tears during a Tobias Wolff reading. I had sat quietly enthralled through passages from his memoirs, *This Boy's Life* and *Pharaoh's Army*, and his novel *Old School*, but as Wolff intoned the final passages from his story "Bullet in the Brain," I broke the silence of the packed auditorium with a gasp, a sob. I clamped both hands over my mouth, but it was too late; heads were turning. One belonged to my new friend Ben George, who at the time was working for *Tin House* and would go on to become the editor of *Ecotone* and Lookout Books. Even then he had an air of consequence and gravity about him. He was a couple years older than I was, with thick, black-framed glasses and a bush of brown hair that stood straight up from his scalp. His demeanor was one of calm intelligence and discretion. And there I was beside him, lungs convulsing, cheeks hot and wet, nose making awful snuffling noises. Later that night, over beers, he asked with cautious sympathy what had set me off. "I just really love that story," was the best answer I could give him, and ever since I've been trying to formulate a better one.

One reason, as I knew even then, was that I had been primed for an emotional reaction by the pages Wolff had read from *Old School*:

> I knew that Maupassant, whose stories I loved, had been taken up when young by Flaubert and Turgenev; Faulkner by Sherwood Anderson; Hemingway by Fitzgerald and Pound and Gertrude Stein. All these writers were welcomed by other writers. It seemed to follow that you needed such a welcome, yet before this could happen you somehow, anyhow, had to meet the writer who was to welcome you. My idea of how this worked wasn't low or even practical. I never thought about making connections. My aspirations were mystical. I wanted to receive the laying on of hands that had actually written living stories and poems, hands that had touched the hands of other writers. I wanted to be anointed.[1]

It just so happened that Wolff was my own literary hero, and that he had workshopped a story of mine the day before, so this passage held special resonance for me. I was a young writer on the verge of hopelessness who had just been "taken up," in some small measure, by the giant at the podium, and at that moment he seemed to look into my soul and see how much it meant to me.

But this couldn't have been the sole reason for my emotional catastrophe because the catastrophe didn't come right then. I listened calmly for another ten or fifteen minutes, until Wolff was concluding "Bullet in the Brain." The story is about a book critic, Anders, who is shot in the head during a bank robbery and, in the moment before death, recalls a scene from his youth. It's a gut-punch of a story, one that shows the wonder and sense of beauty Anders has lost and then gives it back to him, and that must be another reason for my reaction. I really *do* love that story. But in the many dozens of times I'd read it, it had never made me bawl, not even when I'd listened to Wolff read it on audiobook as I drove the high desert of Eastern Washington to spend Thanksgiving with a friend. So why now?

It was the final scene, Anders's memory from his youth, that set me off: "This is what he remembered. Heat. A baseball field. Yellow grass, the whirr of insects, himself leaning against a tree as the boys of the neighborhood gather for a pickup game." Half a page later, the story ends with the passage that brought me to a fever pitch: "For now Anders can still make time. Time for the shadows to lengthen on the field, time for the tethered dog to bark at the flying ball, time for the boy in right field to smack his sweat blackened mitt and softly chant, *They is, they is, they is*."[2] These passages by themselves seem innocuous enough. Each offers a series of descriptions, nothing more. But the conclusion I've come to over the intervening years is that the *description* is exactly what produced my reaction.

And this is important. A common attitude I see from my own writing students these days is that description doesn't serve any vital function in a story. Usually this attitude comes either from laziness—they don't want to put in the imaginative work that produces vivid description—or from a fear of boring readers. As Jerome Stern wrote in *Making Shapely Fiction*, "Young readers often think of description as the parts that they can skip."[3] And so young writers are often eager to do that favor for them.

By "description" I mean the concrete, the things we can observe with our five senses: sight, sound, smell, taste, and touch. I do not mean simple adjectives. I do not mean descriptions such as, "The weather was glorious." Glory is an abstraction, a category of word that George Orwell calls meaningless.[4] By itself, the word "glorious" is useless because it can't show us anything concrete. It can't show a white-hot sun perched overhead, or a sky so hard and blue that a fly ball might shatter it. It can't show a pitcher's shadow puddled under his cleats, or heat rising from the ground in shimmering corrugation. It can't produce the smell of hot aluminum bleachers, or the lubricated slide of a sweaty armpit, or a sunburn tightening the skin on the back of your neck. It can't let you taste the sweat on your lip when you go too long between slugs of cold beer. Only concrete description can do that. As Richard Bausch advises,

> Make your feeling in THINGS, images. There is so much more in an image because that is how we experience the world, and a good story is about EXPERIENCE, not concepts and certainly not abstractions. The abstractions are always finally empty and dull no matter how dear they may be to our hearts and no matter how profound we think they must be. . . . So, in revision, get rid of all those places where you are commenting on things, and let the things stand for themselves. BE CLEAR about the details that can be felt on the skin and in the nerves.[5]

To put it simply, stories should be aimed not at our heads but at our hearts. And this is where things get interesting, because description actually allows access to our hearts in neuro-physical way. Kenneth Kaye, a creative writer who spent forty years in developmental psychology, has written that in a simplified version of the brain, there are two basic regions: the higher and the lower. The higher region is in charge of secondary processes such as facts, math, logic, language rules, moral judgment—in other words, conscious thinking. The lower region is responsible for primary processes such as sensory input, movement, selective memory, and emotional responses—the more automatic systems. These physical locations are part of the reason it seems that thinking occurs in the head and emotion in the body—dread makes your stomach drop, grief squeezes your lungs—

which is, in turn, the reason we refer to the heart, not the head, as our emotional center.

The interesting part is that, as Kaye says, "One can manipulate another person's responses to be less or more emotional by calling upon either the secondary or primary processes, respectively."[6] That means that if you want to defuse someone's emotion—say, the panicked victim of a car accident—you should engage the brain's higher region. You might ask what the date it is, or who the president was fifteen years ago. That requires the person to recall facts and perform computations, activities that are safely isolated from the brain's emotional center. But if you want to intensify someone's emotion—say, a reader—you should engage the lower region. The question is how. The most basic element of a story is language, and language is itself a secondary process that involves the brain's higher region.

It turns out, however, that the brain is remarkably adept at rerouting information to the proper areas for processing. This was first demonstrated in 1969, when a neurophysiologist named Paul Bach-y-Rita unveiled a device that transformed images into tactile sensations for the blind. His device was a dentist's chair with four hundred vibrating stimulators arranged on the back like pixels on a screen. The stimulators were wired to a television camera and a computer that converted images into vibrations. After time, the congenitally blind subjects who used the chair—people who had been blind since birth—were able to recognize objects and people in front of the camera with incredible detail. Even more incredibly, they "began to experience the space in front of them as three-dimensional," according to *The Brain That Changes Itself* by Norman Doidge, M. D. "Their mental perceptual experience took place not on the skin surface but in the world."[7]

This can happen because all sensory information, whether it comes from the eyes or skin or other organs, must be converted into an electrical pattern that can travel through the nervous system to the appropriate area of the brain. For the blind subjects in Bach-y-Rita's chair, the brain had learned to reroute these electrical signals to the visual cortex, the part of the brain that decodes them and produces pictures. It didn't matter that the signals had come from a different "data port," so to speak, as long as they landed in the brain area devoted to sight. As Bach-y-Rita said, "We see with our brains, not our eyes."[8]

So if the brain is capable of rerouting tactile information to the visual cortex, why wouldn't it also be able to reroute information acquired by reading descriptions? As it turns out, that is exactly what seems to happen. When we read, it's not just the language part of the brain that lights up. The *New York Times* has recently reported a growing number of studies that are using brain scans to examine the neurological impact of stories:

> Words like "lavender," "cinnamon" and "soap" elicit a response not only from the language-processing areas of our brains, but also those devoted to dealing with smells. . . . When subjects looked at the words "perfume" and "coffee," their primary olfactory cortex lit up; when they saw the words "chair" and "key," this region remained dark. . . . The brain, it seems, does not make much of a distinction between reading about an experience and encountering it in real life; in each case, the same neurological regions are stimulated.[9]

In other words, when we read about an odor (or image or sensation), it engages the exact same part of the brain as actually smelling it (or seeing it or touching it), and those parts of the brain reside in the lower region, alongside our emotional centers. They're so close, in fact, that there's a high degree of interplay between the two areas, meaning activity in one area can stimulate activity in the other. When you write smells, or images, or sensations, you're actually gaining access to the emotional area of the brain, and this is why stories can take such precise aim at the heart.

Hemingway must have intuited this. Why else would he have known to look for places, in revision, where he'd written statements about emotion and then replace or supplement those statements with images? A statement about emotion—"the outfielder was dejected"—offers only abstraction, and so it lands in the upper brain, where there is no apparatus to stir the reader's emotion. An image, on the other hand, like the ones that obsessed Hemingway—"an outfielder tossed his glove without looking back to where it fell"—dives into the lower brain, where it can trigger in the reader an actual feeling of dejection.[10] In this way, description operates the same way that Kaye says good therapy does: "as if it were a catheter plunged into the limbic system."

And this is, I think, exactly what happened inside me when Wolff read "Bullet in the Brain." Although the earlier passage about writers anointing other writers had moved me, it contained only ideas, and so, poignant as it had been, it was confined to the more intellectual areas of my brain, safely sealed off from my emotions. The end of "Bullet in the Brain," on the other hand, is laden with the sensory description that is capable of accessing the brain's lower region. All those poignant ideas surged toward my emotional center and overwhelmed it, and the result was my public fit of weeping.

But it doesn't end there. These ideas also have interesting implications for how the brain constructs symbols. As Flannery O'Connor rightly pointed out, "symbol" can be a dangerous word.[11] Many people tend to think of symbolism as a kind of mental puzzle, something that requires the logical processing of the brain's upper region, when, in fact, symbolism is what allows a story to build weight in the brain's lower region, orchestrating emotional detonations so complex and powerful that they leave us stunned.

Superstitions come from a disparity between the way the upper and lower regions of the brain process information, according to research conducted by psychologists Jane L. Risen and Thomas Gilovich. The lower region can do it "quickly and effortlessly . . . based on associative connections," whereas the higher region uses a slower, "deliberate and effortful" system of reasoning.[12] You use your upper brain to think things over, but when you have a gut feeling, an intuition, a sense that you know something without knowing *how* you know it, that's the lower region at work. Superstitions arise when the lower region's associations outpace or outmuscle the higher region's logic, a fairly common occurrence. If your favorite baseball team starts scoring runs when you go to the kitchen, for example, you form an association between the scoring and the kitchen, one that can be so strong it might convince you to spend the rest of the game leaning on the fridge, no matter how illogical it seems.

The same associative processing is at work when we read descriptions. As our upper brains are busy converting language into meaning, our lower brains are forming quick and automatic associations between the thing being described and the nature of the description. To say that a man's hair is the color of fine cigars, or dung-colored, or grizzly bear brown, for example, produces no real difference in hue, but it does produce a vast difference in the association it creates. Those associations can infuse objects and actions with deeper meaning, making them symbols. In fact, they can load so much meaning onto a certain object that when something happens to the object, it sets off a choreography of associative reactions in the brain's lower region.

Examples are everywhere, but the best one I can think of is a short unpublished memoir by a friend of mine, Roby Conner, called "On Smoking." The piece is devoted mostly to describing the experience of smoking throughout various stages of his life, and these descriptions load numerous associations onto his cigarettes. It's an exercise in symbolism. When we see Connor smoking as a child, for instance, emulating his father, he writes that "the burn was too harsh for my tiny lungs, but I certainly enjoyed the act: sucking on the filter to make the tip glow, blowing out a plume of smoke that lingered and drifted, flicking the ash." A few sentences later he recalls witnessing his father's ritual with his cigarettes: "He'd hold the pack to his lips and draw one out, flick open his Zippo, strike the wheel and set flame to the end. I told him once, when I was very young, how much I liked that smell, the sharp marriage of lighter fluid and burning tobacco that bloomed from that first puff."[13]

There is certainly more than just description here, but notice all the concrete details—the burn in the lungs, the glowing cherry, the drifting smoke plumes, the blooming odor—and how they establish a link between Conner's smoking and his admiration for his father. That link persists as the piece continues, but Conner complicates it by adding other associations. He goes on to describe smoking as a rebellion, an escape, a muse, a reward, an expression of self-hatred. He describes his dexterity in handling cigarettes, and how rare it is for him to feel such physical grace. When he visits home to see his father, dying of congestive heart failure, he describes smoking "to get away from . . . the hissing goddamned oxygen concentrator that leashed my father to it as surely as a rope around his neck." He describes the shrine he makes from his father's old Zippo lighter after his death. The further he goes, the more associations the reader builds with Conner's cigarettes, and so when he delivers this line—"I've been smoke-free for almost eight weeks now"—we feel an incredible and complex sense of loss.[14] Thanks to the lower brain's associations, we can feel in our guts all the precious things his cigarettes symbolize, and so we are devastated to know what he loses when he throws them away.

To drive this home, Conner returns to the concrete. One of the benefits of quitting is supposed to be regaining one's sense of smell, and so he lets us experience it: "I have, however, been noticing smells more recently. It's spring, so I keep expecting to notice the sweet aroma of blooming flowers. But, no. I smell car exhaust and diesel fumes. My nose is assaulted by perfume and cologne. And shit. I smell shit *everywhere*."[15] This carries a symbolic connotation as well. It's true that he notices more about the world since giving up cigarettes, and he keeps expecting new pleasures, but all it lets him see is how messy and disappointing the world really is. But because this information comes in the form of description rather than in the abstractions I've just used to explain it, it lands in our hearts rather than our brains, and so for one magical moment, we get to feel his disappointment too.

Maybe that's the reason so many readers say that a particularly affective story produces a sensation like the top of the head being "lopped off," as Joshua Bodwell wrote in *Fiction Writers Review*, describing the first time he read Raymond Carver.[16] It's an appropriate analogy. These stories are so successful at saturating the lower region of the brain that they leave the upper region pleasantly vacant. Gone, or at least immaterial, are the logical faculties that get in the way of direct experience, and all that is left is the tingle on the skin, the blow to the gut, the muffled sobs that break an auditorium's kind silence.

10

The Memory Agent

David Lazar

Hannibal Lecter, in *Silence of the Lambs*, says, "Memory, Agent Starling, is what I have instead of a view."

When I was six, we were going on an outing on a cold February day. My brother and mother were in our car, and I was standing at the top of our row house stairs, my father locking the door behind me. There was snow and ice on the brick stairs leading down to the driveway, and as I looked down the stairs through my hood, down the stairs to the car with its puffing exhausting waiting below, I became paralyzed at the thought of the impossible obstacle course of the stairs. The ice and snow, I remember, seemed like a moat, unbreachable, things that I could not pass. My father said, "Walk down," or something wildly encouraging to that effect, but I was utterly immobilized with fear. He started slowly walking down, hand on rail, and turned back to me: "Walk down." I told him I couldn't. I told him it was too icy. He firmly instructed me to hold to the rail and come along. I couldn't. I wouldn't. This put my father out, I suppose, because rather than responding to my fear of slipping and breaking my neck or some such thing, he responded to a perceived disavowal of his command. In short, I was disobeying. He said that if I didn't walk down the stairs, they would leave without me. I informed him of the impossibility of acceding to his demands, though certainly not in those words. I stayed at the top of the stairs, and my father gingerly climbed into the driver's seat, pulled the smoking Buick back into the street and drove away. I remained, hooded, at the top of the stairs, watching the car, my family, diminish in the distance. I remember looking back down at the stairs. I remember a vague sense of abandonment and timelessness and I stood there in the cold. It couldn't have been for terribly long.

The Buick came back down the other end of the street and pulled into the driveway. My father, furious (At what, I think now—did he expect me to walk down the icy stairs in their absence? Or was he mad about coming back for me?), got out of the car, climbed briskly up the stairs, and carried me down to the car.

I tell this story because in the approximately fifty-five years since I have been unable to walk on snow or ice, thin or thick, without wobbling with a sense of imminent danger. Especially ice: My frozen moment froze something in my body, which it remembered, does remember still, so that habit and trauma have combined to make me an tentative walker in the cold, my hands out like I'm walking on a tightrope.

My epigraphic source might seem a bit a bit dubious, though I think we are hungry to try and place ourselves in memory in ways that satisfy and make sense.

There is a connection between memory, movies, and dreams of which Delmore Schwartz was famously aware in "In Dreams Begin Responsibilities," his autobiographical story, or story-essay, about watching the film of his parents' courtship on a movie screen in a dream on the cusp of his twenty-first birthday. Movies sometimes seem like dreams, and dreams seem like movies. Our memories sometimes stutter in our minds like projected images of celluloid. The Golden Age of cinema in Hollywood was called the Dream Factory, and sometimes when we try to piece together memories it is as though we are editing a film, silent or sound, constructed of scraps of detritus that our cortices have preserved.

In *Matter and Memory*, Henri Bergson writes that "to perceive consists in condensing enormous periods of an infinitely diluted existence into a few more differentiated moments of an intenser life, and in thus summing up a very long history" (273). He captures the combined and mystifying strands of perception and cognition that scientists and psychologists have struggled to weave into a unified theory of memory, instead giving us glimpses for the most part of how the mysterious inner vision of memory works. Memory is a combined effort of various parts of the brain, and Bergson felt that we were only truly, flamingly alive when we became conscious of the workings of memory, combined with our experience of the sensual world in the moment—Bergsonian moments that became transformed in the work of Woolf, Joyce, Eliot and many others. Forster's room with a view is expansive, not really so different, startlingly, from Hannibal Lecter's, in that the mind has to go in, to memory, in order to go out to the world. In other words, we are in many ways what we remember, though how we remember seems like a scientific Tantalus, always seeming just out of reach even as we know more and more.

* * *

In a major study of perceptions of how memory work *What People Believe about How Memory Works: A Representative Survey of the U.S. Population* by Daniel J. Simons and Christopher F. Chabris, respondents (a large sampling, over fifty thousand) agreed to the statement in the survey that "human memory works like a video camera, accurately recording the events we see and hear so that we can review and inspect them later" by an extraordinary 63 percent. Two-thirds of people think they're video cameras, ignoring the flaws, gaps, delusions, desires, physiological deficiencies of age or injury (concussions, dementia) that common memory would seem to present. All sixteen experts, members of the Psychonomic Society (Can I join?) naturally disagreed. People want to trust memory. I felt a bit jealous—I don't even think I'm a pencil, and all these thousands of people think they're cameras.

New research from the University of British Columbia suggests that people's attitudes toward what they believe may be much more malleable than we think. Plastic water bottles were banned in San Francisco and a lot of people were really bothered by this. But just a day after the ban was introduced, researchers found that attitudes started shifting—people who were opposed began changing the nature of their opposition, loosening the memory of why they so did not want the ban. The researchers extrapolated the idea that memory is fungible because we want to be happier. And we alter ideas and memories, frequently, perhaps even usually unconsciously, to make ourselves less anxious and malcontent. We modify, selectively omit, change details, or rewrite to make, well, yes, a happier ending, that's the unconscious desire, but a happier beginning, a brighter earlier stage of memory. Kristin Laurin, author of the UBC study, detailed in *Inaugurating Rationalization: Three Field Studies Find Increased Rationalization When Anticipated Realities Become Current* likens this to a "psychological immune system [. . .] your brain is scrambling to make you feel okay and allow you to get on with your life."

This striving toward a happy story isn't inevitably true. Sometimes, neurotically, we push memory into darker places, to forge narratives that serve a fundamentally different purpose, to engorge our pain, which may serve us in different ways.

But for most people, memory isn't a set of questions, but rather a set of answers, a file. Each age, each epoch, has its distinct metaphors for memory, depending on the distinct set of cultural constructs and technologies available. "Metaphors of memory are overwhelmingly physical [. . .] we use verbs such as impress, burn and imprint to describe the processes by which memories are formed," Charles Fernyhough reminds us in *Pieces of Light: How the New Science of Memory Illuminates the Stories We Tell About Our Pasts* and Douwe Draaisma in *Metaphors of Memory: A History of Ideas About the Mind* devotes himself to this idea. Our belief about human memory seems to be one that is repeatable, crisp, generally available, and speaks to us in unproblematized ways.

Memory perhaps has just become meta-memory since reality is a form of reality TV. In other words, we are living memories as we make them, seeing ourselves and filming ourselves. So why shouldn't most people see memory itself as an extension of the film of themselves they are shooting at the moment, on their phone, tablet, and so on.

In the Simons and Chambris study, 47.6 percent of respondents agreed with the response option that "once you have experienced an event and formed a memory of it, that memory does not change." Half of people questioned, in other words, believe that memory is permanent, fixed. Like the song says, "It's Easy to Remember (And So Hard to Forget)." How is it, one wonders, that "I don't recall" plays out across our political landscape like a perseverative refrain, a get out of jail free card for blithe memory loss in the midst of a culture that seems to believe that memory's memory banks are fairly fixed.

Of course, we're brilliant and creative translators of our lives, not machines. Memory is the ultimate essay, based on distant fact, requiring imagination, flawed and inflamed by desire, crippled by its partiality, necessary to a sense of identity, most useful when approached with irony and self-doubt, and informed by, crucially, a sense of the real body's existence in time.

What is called *body memory* designates the pre-thematic impact of experiences on the meaningful, and yet implicit, configuration of our conscious experience registered on the body. In other words, the body, as entity, has its own system of awareness over time, distinct from the figurations of cognitive conceptions of memory. Body memory embraces the totality of our subjective perceptual and behavioral dispositions, a specifically material mediation of the self.

Edmund Husserl, generally considered the founder of phenomenology, shows that body memory plays a role in both perceptual constitution, the reality of what is visually understood, and the formation of bodily, pre-reflective self-awareness, our sense of the body before the sense of a self. In both cases, body memory allows us to get acquainted with perceptual and experiential patterns, and to acquire familiarity with our bodily capacities. Moreover, body memory is responsible of the formation of an individual style of perceiving and moving, and, more generally, of experiencing the world. As such, body memory is the condition for the formation of the bodily "I can." Our bodies learn to operate and move in the world through nerves, through repetition, through what we call instinct. Bergson speaks of "memoire-habitude," the habitual memory that the body acquires, beginning very early, though it can be developed, even refined. Maurice Merleau-Ponty points to the work of typists and dancers, and we might add acrobats and gymnasts, you and I, the ways bodies learn through repetition to move with their own volition.

Edward Casey distinguishes between three forms of body memory: habitual, traumatic, and erotic. The categories can overlap. In doing so, Casey stresses that "body memory" cannot be simply equated with the "memory of the body," if by the latter the body is conceived as the correlative of acts of memory. Rather, body memory in its different forms is considered to be "intrinsic to the body, to its own ways of remembering: how we remember in and by and through the body." The body remembers differently: It doesn't think memory, instead it does something closer to *being* or *enacting* memory.

On a note related to Casey's ideas on body memory, Thomas Fuchs, the Karl Jaspers Professor for Philosophy and Psychiatry at Heidelberg, and his associates propose a descriptive taxonomy that includes six forms of body memory:

Habitual or procedural memories are acquired sensory-motor skills and attitudes that get us acquainted with perceptual and movement patterns. Through our body's habitual memory we develop an individual style of interaction with the world, the way we walk, for example, and our gestures.

Situational memory "makes possible both the feeling of familiarity with some situations and the feeling of alienness with respect to others," which allows our bodies to relax into or withdraw from certain situations. This is literally the body in situ, in places where it feels comfortable or ill at ease—how do you hold yourself in a class, on a doctor's table, in front of someone who is more or less powerful than you? Sabine Koch, Thomas Fuchs and Mimmchela Summa talk about this in "Body Memory and Kinesthetic Body Feedback: the impact of light vs. strong movement on affect and cognition."

Intercorporeal memory is learned early and interiorized, incorporated, and mediated by other experiences, based on imitation and identification, and according to the authors, in their Lacanian view, it is based on the acknowledgment of the gaze of the other, a sense of the self's body as distinct. How infants are held and carried is an embodied memory which translates into later feelings about intimacy, bodily proximity, the self.

Pain memory is, naturally enough, the lingering effect of painfully body experiences which linger over time. If, for example, you walked into a glass door and fell flat on your face, you would be cautious in buildings with lots of glass surfaces, extending your arms, maybe looking a little like a sleepwalker.

In *traumatic memory*, extremely painful experiences linger on in their bodily manifestations disconnected from a consciousness of their original source. The memories sources in a body's movements, gestures, postures, can revive the pain of the trauma. Sex can revive memory of sexual abuse; a tilt of the head can suggest the memory of a traumatic bit of tormenting information overheard years earlier, again without the explicit memory of the event, though this can, too, sometimes be triggered.

There have been different taxonomies—perhaps you have your own, your sense of how the body remembers and forgets. And lest we fall into a Cartesian malaise or miasma in reverse, it's important to remember that considering the body phenomenologically and not integratively, while suggestive, threatens to leave us Ichabod Craning: trying to figure out what to do with the headless horsemen of ourselves and forgetting that neuroscience and cognitive science and phenomenology are still scientific metaphor-generators for thinking about how we think: Montaigne writes that "nothing fixes a thing so intensely in the memory as the desire to forget it." And so when we turn to write, when we write about memory, about bodies, about the memory of our bodies, in essays that try to deconstruct the memory machine, it's worth noting the complex immutability of the law of desire: what we want is not necessarily what we remember.

* * *

"All my people are larger bodies than mine," James Agee writes in "Knoxville, Summer 1915," but *I* felt, as a child, that all of *my* body was larger than other people's, and this largeness was shapeless, grotesqueness, and blameful. From about six to fourteen, I swelled and experienced myself as misshapen and gluttonous. Before the epoch of obesity, I was merely fat, or not so merely, since while I experienced external misery in the forms of mockery and disgust, from family, schoolmates, an interiorized projection of the world's very judgment itself, as though God, whom I had a rather vague connection to, were looking at me, outside and in, and saying, *Why are you like that, and why do you look as you do?*

If we use Fuchs's taxonomy, from above, we can see that my misery was tapping into malignant body consciousness of several different kinds that would last, well, a lifetime. My "pain memory" is present every time I look in the mirror and see a body that is less svelte than I would like; I remember the gruesome self-disgust of the thirteen-year-old boy, body, boy body, gazing in the mirror with the appropriated self-hatred of a swollen self-image of distended disdain.

And this is Fuchs's traumatic memory, as well, having impinged on the experiences of my past forty-five years, sometimes, as he suggests, without any awareness of the damage that was wrought early on.

And my incorporative memories, the incorporation of the corpus, well, the decades of behaviors neurotically protecting an injured psyche, working out or merely petting the scabs of early years experiencing a self in full-blown horror. Is this our thrall at horror movies? A kind of Lacanian expression of self-loathing, permanently stalled? "This is no dream," Rosemary says in *Rosemary's Baby*, "this is really happening." Bad dream, bad memory, frozen gaze of the self trying to speak but silenced or muffled by repressed pain. My larger body, Delmore Schwartz's "heavy bear" (I even named my son Delmore after a man whose life was full of pain.) is still with me, this body memory that "dressed in his dress-suit, bulging his pants, / Trembles to think that his quivering meat / Must finally wince to nothing at all" as Schwartz writes in "The Heavy Bear Who Goes With Me."

Generally, the "reconstructive account of memory [. . .] is largely accepted by memory scientists." One doesn't, contrary to popular belief, turn on the video camera of one's mind, but creates something new. Daniel Schacter, in *Searching for Memory: The Brain, the Mind and the Past*, writes that we "re-create or reconstruct our experiences rather than retrieve copies of them. Sometimes, in the process of reconstructing, we add on feelings, beliefs, or even knowledge we obtained after the experience. In other words, we bias our memories of the past by attributing to them emotions of knowledge we acquired after the event." For those of us who write autobiographically and teach autobiography in one form or another, this is hardly news, as I'm sure is true for anyone who has thought hard about their memories. Our thoughts are inflections—impure and variable currents of desire. This doesn't mean that all memories are pure inventions—things did happen, history avers. But as is the case with history, after the act comes the shadow of interpretation, sometimes well-lit, sometimes not, and the challenge is to try and understand both what we see in a given light and what the light means.

Elizabeth Loftus, an American cognitive psychologist, has for over thirty years conducted experiments on "false memory," participants who could be led to believe something had happened to them when it hadn't. This led her to challenge certain claims of "repressed memory," a highly fraught subject in the memory field, and an area where feminist writers and researchers are mindful of the dismissed claims of retrieved trauma (other memory researchers, such as Steve Ceci of Cornell University, argue for the concept of "motivated forgetting," a simpler concept based on the idea that we forget what is painful, what we'd rather not remember, but that these memories are not necessarily out of reach, that they still may be accessible through semantic or other links). But Loftus goes further, suggesting implanted memory can be useful for curbing the traumas of PTSD, for example, or in eating disorders. Queen of Hearts, anyone? I immediately think of *The Manchurian Candidate.*

The work of Elizabeth Loftus is in a line of scientific research on autobiographical memory, some of it quite contradictory, that dates back a 150 years, to the work of Francis Galton and Hermann Ebbinghaus. But the first proper work on memory as reconstruction was performed by the first professor of experimental psychology at the University of Cambridge, Sir Frederic Bartlett. Using "The War of the Ghosts," a Native American narrative, he tested his subjects recall and determined how narratives were pieced together, how they were remembered with additives of personal ephemera and detritus, things that the subjects wanted to remember and add to the memory, and things that they wished to leave out. As Charles Fernyhough notes in *Pieces of Light*, Martin "Conway has distinguished between two forces in human memory: the force of correspondence, which captures memory's need to stay true to the facts of what happened, and the force of coherence, which works to make memory consistent with our current goals and our images and beliefs about our own selves" (12). In other words, remembering is always engaged in a creative tension, between fact/event and desire/construction. Some of creative tension is conscious and some is unconscious, as is the case with the creation of art. Or even more specifically, let's say essay writing. We want to try to hold to a fact, to think about it, to internally or even externally test what the facts of an event may have been. But what we want from the past is complex and manifold, though we might not want a past that is manifold. We struggle to find a version that makes sense and that satisfies our impulse to rethink it. We think the images in our mind, subtly altering details. We try to coordinate what scientists call semantic memory (memory for facts) and episodic memory (memory for events). And again, in this midst of all of these swirls around each other what are called declarative memory or explicit memory (those we are aware of, conscious of) and implicit or non-declarative memory (the unconscious).

Memory: *From Brooklyn*

I was about twenty years old, and I always want to remember this memory as young as possible. So when I think about this memory without questioning the details (the implicit memory) I'm perhaps eighteen or nineteen years old, wandering into Greenwich Village on a Friday night, looking, as they say, for trouble. But I must have been older. My semantic or factual memory, exploiting the demands of my explicit memory to interrogate what is actually known and what I've assumed or even ordained, pushes me *think*: I couldn't have been that young on that night I went into the Village to drink, to try to lose a bit of the darkened self that seemed to plague me in my late teens, early twenties. Oh hell, for many years. And, perhaps, I thought, I might meet someone, find some distraction. But I must have been a little older, twenty-one perhaps, which makes me a little more ashamed of what happened, since I was less young, and therefore am less, in memory, able to write things off as casualties of "callow youth." Even a couple of years can soften the blows of conscience.

So I was headed to Manhattan, and I'll tell you it was from Brooklyn. I haven't seen a category of memory for parts of our narrative that are undergirths, or substructures, those elements that we take so much for granted that we don't even have to consciously register: such as *from Brooklyn*. I spent my entire childhood in Brooklyn, so that when I remember sledding down a hill, or wading into the ocean, hearing the El at night or watching a film with Maria Ouspenskya, masturbating in my downstairs bedroom on Ocean Parkway, or eating chicken soup with one big carrot in it, it's in Brooklyn, the geographical girding of the memory that's so ever-present it usually goes unremarked. It's the primer of my early memory canvases, and note the use of a static and very early metaphor for memory: painting as memory metaphor first showed up hundreds of years ago.

So I was headed to Manhattan and I'm sure I wanted to drink and I'm sure I wanted to meet someone. These semantic assertions are based on the inchoate but very real pain memories of feelings—the years of barnacled melancholia I thought were just part of the constant drift. As a young man I didn't know I could or would ever feel differently, so I sought, not irrationally, distraction. I really did drink a lot, and like many of my friends I sometimes marvel that I survived a host of absurdly dangerous and carelessly threatening situations.

Better to have been younger, I have always implicitly remembered thinking of these times.

I went to various bars in the West Village, which in the mid-1970s was, like much of Manhattan, grittier, dirtier, edgier, and entirely more interesting, at least to me now. The me of eighteen or twenty-one would agree, but only because I insist upon coherence—the young man was with the aesthetic program. Good boy.

In one bar, after several bars, let's call it the Village Bar since I'm sure it doesn't exist anymore. (You see the way I insist it doesn't exist anymore without any information, so it will be of a piece with my sense of the changing city, the lost self, the demands of narrative cohesion.) I walked in, slight woozy, having had a few of whatever it was I was drinking at the time: White Russians, Jamesons, Black Russians, Rusty Nails? I can sometimes almost remember who I was by thinking of the smell (the olfactory being the strongest memory sense) and taste of what I drank.

In any case, I walked in and I must have gone up to the bar to order. In looking at research on memory, I haven't found a name for the filling in we do, the interstitial memory clay that connects what we remember to what we remember. You remember, for example, talking to your father, turning, ten minutes later the accident, the hospital, then . . . . In telling the story do you elide, jump cut (We use movies as a most common metaphor for memory.), explain the absences, or create a fluid movement, filling in as needed, again, the cohesion narrative mentioned above? If you do fill in, there is a chance you're creating a kind of false narrative to retell later as remembered truth, since we remember the versions of memory we have remembered as though they were original, biblical, and not a form of telephone we have played with ourselves over the years, whispering the story in ourselves, only sometimes to repeat, and always different from the first version, how much a question of degree.

So I must have gone up to the bar to order. And after I do that, drink in hand, I somehow ended up at a table with a woman with a wooden leg.

I remember her as very pretty and older than I, perhaps thirty-five. And she must have liked me in some way, because I ended up sitting next to her, and I remember our conversation as tart and engaging, but this is where the fringes of memory start to fray (memory as curtain, to pull aside or close? Memory as coverlet?) since I was drinking, lots. Whether it was White Russian, Black Russian or Red Cavalry, all I can really remember from this point on in the bar is that I had my head resting on her shoulder because I was very drunk (and because I was sexually attracted to her, frissoned by the age difference and her confidence, and yes, no doubt, what seemed the strangeness at the time of a wooden leg) and there was a very small man, suddenly sitting opposite us, whom I felt was a rival for her attentions. But this part of the memory seems most unreliable to me. I can say with complete confidence that I was in a bar and was talking to a woman with a wooden leg—these are semantic and explicit memories that I am sure of, that do not waver, that remain ensconced in the Hall of

Facts that do not need to be dusted off and carbon dated, or shown to my team of memory testers for authenticity or veracity. Good to go, for telling and thinking, my inner team says! Seal of factual approval! Good Memory Housekeeping. But the memory of the little man never gets the seal (a sentence I hope you find as strange as I do).

Endel Tulving, a psychologist and cognitive neuroscientist, called our ability to inhabit memory "autonoetic consciousness." We know that the hippocampus is related to the spatial and psychological processes of memory, the occipital cortex to the visual processing, the amygdala—that small mediating corpus—to emotional associations and retentions, but the experience of memory, the inhabitation of the past as a series of personal moments, subjectively, remains elusive.

I, for example, remember a past moment, clouded by sadness and inebriation with a small man looking at me across the table. Was he there? Or a projection of my own diminished ego, added in the years that followed as I remembered the moment and then remembered the remembered moments, narrative ripples. The memory is what I decide to think. And several different versions may juxtapose. Many people think they have static memories, forgetting that the good day or rotten day can determine both the feeling of what we remember on that day and how we arrange the details. *That's not the way you told it last time*, we say to the rememberer, as though memories were committed to memory. They aren't. They're images, of varying intensity, that have stuck around our cortexes, in forms we don't fully understand.

I fell asleep on the woman's shoulder, and she left, with or without the little man who may or may not have been my rival. I had drunk too much. Way too much. And I stumbled out into the night, really early morning, too drunk to make my way back to Brooklyn. It was, I think—this is the kind of conjecture that makes up large parts of memory the way that dark matter fills the universe—all that's there but that can't be seen—around three or four in the morning. I made my way down to West 3rd Street, I'm almost sure it was West 3rd Street, and found what seemed to be a reasonable doorway, meaning it was open, dark, and not particularly foul.

If you had been there, you might tell me there was no other man, no mediating presence for my young melancholic warp and woof. If you had been there, you might argue that the guy sitting opposite was about my own height. If you had been there you might have shaken your head and left just as I was sitting down, and would say to me now, *I can only give you a partial version from point of view, because I was driven, that night, by my own set of circumstance, exigence, and you are really a minor character in my episodic memory. Sorry.* This acknowledgment would give your version a jostle of veracity. But you would be obeying, as the experts say, your own synaptic and cortexial impulses to construct. You would be telling the story that you needed to tell, wanted to tell, and felt you could only tell. But the important point is that you would be writing an autobiographical narrative, narrating an essay, blending the feelings that still reverberate in your brain through the complex distributions of the limbic system in the hypothalmus, the cingulate gyrus, the hippocampus, and again, the amygdala, that mysterious island. You'd combine the feelings, a hot core of which might resemble what you first felt (Or might not? Remember, we change our memories as we need.) and dress it on the sense of what you did and saw. Perhaps as we compared our versions of what we remembered, we would marvel at the idea that the past is implacably nonexistent except as we think it, except as memory.

* * *

L. P. Hartley, famously, wrote, "The past is a foreign country. They do things differently there" (*The Go-Between*, 1953). The science of memory shows us that our brains combine different and complex functions for this time travel. Sometimes we land in a dark night with women with wooden legs and men who flicker in the shadows as either real or apparitions of ourselves. Montaigne knew the self to be shifting and contradictory, and scientists and theorists of the self, neurologists and psychologists, philosophers and linguists, from William James to Hermann Ebbinghaus to Oliver Sachs, understand that the self is both a pattern of behavior, an individual's own sense of what we might call the self-state over time, and the shifting pool of memories that the self retains, consciously and unconsciously, and sees itself an agent in. What we remember is to a large extent who we are. To forget, I think (as I wonder why I was so unhappy that night so many years ago, why I lay my head on the shoulder of a woman I didn't know, why I hoped she could make me forget whatever I needed to escape), is to lose some part of myself and crucially not remembering moments and scenarios, images and sensations.

Sometimes that is what I want, don't you?

11

On Metaphor[1]

V. Efua Prince

Like story, metaphor conveys. The word "convey" is constructed by joining the Medieval Latin *con*, which means "together," with the Latin *via*, meaning "way." The image evoked by conveyance is spatial, suggesting movement from one place to another. If metaphor conveys, then metaphor is a means of bringing something across. Moreover, if the image suggested by the verb "to convey" is spatial, then it also evokes a Newtonian idea of time—which is to say that crossing a span necessarily takes time. The time that it takes for metaphor to convey a thing across whatever given distance is one of the characteristics that distinguishes metaphor from story and other forms of linguistic expression. The fact is that metaphor conveys with remarkable efficiency. Its efficiency results from three primary factors: (1) metaphor's ability to unfold in a moment, (2) its ability to generate tension, and (3) its ability to call a thing into focus.

Metaphorical equation

On the most rudimentary level, metaphor is a comparison without the use of like or as. Metaphor might be rendered as this simple equation: $A = B$. Metaphors work, however, precisely because $A \neq B$. The comparison of A and B brings similarities between dissimilar things into focus. Metaphor purposefully applies a thing generally understood to be a part of one category with characteristics of another category. In short, as Richard H. Brown describes them, metaphors are "category errors with a purpose, linguistic madness with a method" (82). Metaphor's method is to create new associations and new categories by deliberately assigning things to the "wrong" category. For the purposes of this examination, however, I situate metaphor under a larger notion of metaphorical thinking, which is most concerned with the science of comparison. In this way any number of devices such as similes, contranyms, puns, hyperbole, and analogies, in addition to expressions that may not be identified as a linguistic device, operate metaphorically because the focus here is on the comparison, which may or may not be overt.

Take, for instance, a sentence from Toni Morrison's *The Bluest Eye*: "The pieces of Cholly's life could become coherent only in the head of a musician" (159). This sentence is built on the metaphorical equation life = pieces, which is to say that what one understands as a person's life—in this case Cholly's—is fragmented. Rather than an integrated whole suggested in the singular notion of a life, Cholly's life is presented as fractured, which when taken together creates a whole. But the metaphor goes on to qualify the code assigned to those pieces. Thus, the basic equation is modified: life = (in)coherent pieces. Most people cannot easily reassemble the pieces of Cholly's life. The fragments alluded to here are fractions that require a particular kind of reasoning in order for them to be understood as a whole life. So Morrison adopts a simple metaphor: life = pieces (which most people would readily accept and hardly notice as metaphorical) and adapts it by qualifying the terms under which one can understand the premise of the initial metaphor.

The delight that readers experience when encountering a sentence like this comes from the modification Morrison makes to a fairly common metaphor. Essentially, she awakens a dormant metaphor (life = pieces) by putting the language back into play. The play occurs when Morrison introduces limits to the established set. So the reader encountering a familiar metaphor confronts unfamiliar provisos when a variable is added to the equation (life = coherent pieces *when x*; with *x* given as "in the head of a musician"). Thus the metaphor disrupts reader stasis by moving the conceptual frame. It animates the language, transporting the reader from the common to the delightfully fresh. In this way, Morrison conveys a remarkable amount of information, not so much about the character Cholly, as about the mind that is incapable of reasoning through the "pieces" which make up his life. She tells us that there are people who have the potential of interpreting Cholly's life as a coherent whole. Those are musicians—a significantly smaller subset of people. All the rest, then, find what is given as fragments of Cholly's life to be incoherent.

The novel from which this sentence is drawn tells the story of Cholly's daughter's descent into madness due to her belief that she is ugly because she is black. Cholly is implicated in both her blackness and her descent. It is critical, in such a story, to have mechanisms for unpacking language, without which the language itself works to reinforce the very destructive paradigms Morrison hopes to deconstruct. By implicating the mind in which the coherence is formulated in the metaphorical equation given to bring insight into Cholly's life, Morrison, in effect, distinguishes the vast majority of her readers from the subset of individuals who will be able to read Cholly's life as an integrated whole. In this way, Morrison moves Cholly into a safer space within the literary imagination that is generally beyond the reach of dominant codes. The deficit, if someone must be held responsible for what happens in the course of the novel, is in the mind that cannot make the life coherent rather than in any particular failure of the life in question. Thus the code that sets up the expectation of coherence, which is to say the norms that one applies to understand what a life is supposed to be like, is disrupted. Without the tyranny of codes, readers are able to grapple with the limits of their own interpretive possibilities. Such a shift is crucial. That Morrison is able to accomplish so much of this within the span of a single sentence suggests the efficiency of metaphor.

Metaphor unfolds in a moment

Andrew Chekhov offered this advice about storytelling: If in the first act you have hung a pistol on the wall, then in the following one it should be fired. Otherwise don't put it there.[2] Clearly, Chekhov is concerned about superfluous detail. He wants playwrights to avoid distracting the audience with needless detail that does not relate to the development of the drama. But there's another aspect to Chekhov's advice. Chekhov's gun speaks to an audience's interpretive expectations. An audience finding a pistol on the wall will read the gun as part of their interpretive engagement and so will expect it to have later significance. The placement of a weapon on a wall should be deliberate. Conversely, if a gun is to be used in the second act, the audience must already be aware of the presence of a gun. The earlier awareness lays a foundation for a later event. A narrative necessarily takes time to unfold.

Chekhov offers cogent advice because he is a master storyteller, which is to say that he understands the conventions that shape audience expectation. Story operates under a set of codes that signal the encounter with a story. One of the most familiar models is the Freytag pyramid. The Freytag pyramid presents a narrative arc as an introduction, a problem, rising action, climax, falling action, and dénouement. It is a common model for Western stories because, as Brian Upton, a game designer who studies the way play intersects with narrative, argues, the Freytag pyramid "is a particularly useful way to establish a robust and flexible play space" (257). As is the case with metaphor, interpretive play in a story is an elastic space opened up by language, which encourages conceptual leaps and imaginative connections. Upton explains that "in order for a story to function as a story, we need to feel as if we understand where it is headed. We make interpretive moves that structure chains of anticipatory play. Because of this anticipatory play, we have a sense of what might happen and what will not" (246). Evidence offers clues to upcoming events in order to create the tension necessary for an audience to recognize what ties the parts of the narrative together as a story. The framework is crafted for its ability to convey any number of messages from the storyteller to an audience. Yet while the elasticity of the Freytag pyramid may appear to offer an infinite range of possible stories, it is not, in fact, a neutral, uninterested frame.

In his article, "Narrative Structure and Text Structure," John Holloway describes narrative as a set of "runs" of events. Holloway explains that

> a literary narrative need not be a single sequence of events: on the contrary, most literary narratives are complexes consisting of several sequences. By definition, the sequential relations between the items in each run are determined exhaustively; but the sequential relations between items in the various runs may not be. (583)

Which is to say that narrative may contain any number of runs of events but the number of events within a given sequence is limited; however, narrative may have any number of associations across the given runs. Holloway represents the summation of narrative as "Σ." The idea of narrative as Σ is important because Σ is used as a mathematical symbol to evoke a whole even as it acknowledges the presence of constituent parts. In this case, the parts are runs of events.

Like the word "convey," even when operating as a noun as it is here, the image associated with the word "runs" is spatial. A run takes place over a period of time and across a given distance. In a diagram showing how the events operate within the narrative, Holloway renders the runs that make up Σ as vertical, parallel lines. Holloway later offers a modification to the diagram by rendering the runs as parallel lines drawn around a cylinder. While the initial diagram is useful for demonstrating runs in operation, it is conceptually incomplete. Holloway suggests,

> To emphasize how in each case the final state is the same as the initial one, the diagrams could be drawn on the surface of a cylinder with time represented by a determinate direction around the cylinder. Possibly this is better, because it represents the time of the novel as bounded, and it makes good sense to say that there is no novel time before the beginning of the book or after the end of it. (587)

While it happens in the particular story that Holloway references that the final state of a narrative run is the same as the initial state, it may not necessarily be the case; nevertheless, in imagining the diagram as an arrow wrapping around a cylinder to end at the place where it began, Holloway more clearly conveys two essential facts: (1) a sequence of events happens over time and (2) time within a story is bounded. Both points position the matter of a given length of time as critical to the operation of story.

A story's bounded length of time operates alongside other literary devices "to structure an idealized epistemological playground—a mental space in which we can move easily through our natural cycle of knowledge construction and application" (Upton 258). A story's epistemological playground necessarily consists of a run of events unfolding over a given period of time. Metaphor, on the other hand, may unfold in a moment.

In mathematics, a moment is a specific quantitative measure of the shape of a set of points. The tension between the two categories A and B clarifies which characteristics should be called into focus, thus defining the shape of the set of points. The difference between the categories operates like distance in physics—the greater the difference, the greater the distance. While the image of conveyance suggests linearity, metaphor relies upon a fixed reference point and generates an amount of conceptual energy based upon the distinctiveness between A and B. In this way, metaphors are constructed through conceptual tension. This tension is produced by balancing the right amount of difference between the first term, A—the target—and the second term, B—the source. The ground, which is the amount of resemblance between the two terms, relies upon this tension as a sort of stabilizing force. Too much similarity between the first term, A—the target—and the second term, B—the source results in banal insights. For instance, "the leader is a boss" is a poor metaphor. The ground is weak because it is over saturated. Conversely, too much difference between the first term, A—the target—and the second term, B—the source results in absurdity. For example, "the album is a hamburger" is an utterly useless comparison. The ground in this case is also weak because the connection between the terms, the album and a hamburger is too thin. Good metaphors rest on an adequate amount of similarities between the first term, A—the target—and the second term, B—the source, to envision a correspondence. While the correspondence is necessarily incomplete because the target and the source are not the same, it is complete enough for one to envision a sense of unity.

Imagine the correspondence as a rubber band fixed at one end to the source and at the other to the target. A weak ground, characterized by slack in the rubber band, lacks the energy required to pull the target to the source. Conversely, a strong ground generates energy by stretching the rubber band. With sufficient correspondence, the rubber band stores enough energy to generate a snap, and the distance between A and B can collapse in a moment. At that moment, the specific set identified with the source is applied to the target and refuses to acknowledge any difference outside of the set. Unlike the epistemological play maintained by narrative that unfolds over time, the elastic energy of metaphor is more immediate.

A metaphorical moment

In her now famous comments on "The Personal and the Political Panel" at the Second Sex Conference held in New York on September 29, 1979, Audre Lorde insisted that "the master's tools will never dismantle the master's house" (112). Part of the reason her words continue to resonate is because of the strength of the central metaphor. The master, his house, and his tools are set in opposition to individual creativity enabled by an interdependency between and across women's differences. Through this metaphor, Lorde usurps the role of the master by utilizing her distinct subject position as a black, lesbian woman and by asserting her right to speak. Her words and her critical frame, then, operate in place of his tools. Thus, the metaphorical equation here reveals the illogic of utilizing tools, which have been conceived in the service of the master's house, in a counterintuitive effort to dismantle his authority. Furthermore, within the context of the United States, the master's house necessarily connotes a history of enslavement, white supremacist, patriarchal domination, and a legacy of racialized and sexualized violence—hence the necessity for its dismantling. Nevertheless, because the "tools" are metaphorical and not actual machinery like bull dozers, wrecking balls, hammers, and saws, but are instead ideas, philosophies, systems, and language crafted to work according to a master code, Lorde insists that these tools cannot operate against their design.

If we stay with the mathematical definition of a moment as a specific quantitative measure of the shape of a set of points, then Lorde presents us with a metaphorical moment that is characterized by a set of points associated with the notion of "the master." The master denotes a skilled and erudite overlord with sovereign patriarchal control over a domain. But it also connotes a brutal and illegitimate colonizer who uses force to maintain a tenuous hold over a territory and its peoples. The full range of these points shape the set captured by this metaphorical moment. Lorde unironically calls the master fully into view. She brings to the surface connotations that might otherwise be sublimated by the normative presence of "the master" and what that presence has meant for many in the shaping of society. "The master," in the formulation of Lorde's metaphor, is given to be necessarily different from the diversity of women Lorde offers in his stead.

The foundational equation of Lorde's metaphor is this: the master's house = academy; the master's tools = traditional academic discourse. Here the master and his tools operate as the source (category A), while the academy and its discursive practice operate as the target (category B). The metaphor focuses our attention on the construct of the master and his trappings. This metaphorical moment places the characteristics associated with the master in play. Once they enter the more elastic space of play, they can be conveyed from the source, which in this case is the "master," to another target, which Lorde ultimately constructs as a diverse womanhood with their own domain and set of deconstructive tools. The difference between what one conceives of as "the master" and Lorde's representation of women creates tension. This tension begins in formulations of race and gender and class that extend beyond the academic. While the academic seems to be a hermetic system, Lorde looks outside the "house" for solutions that necessarily involve opening the apparently closed system. As Lorde explains, "Know that *survival is not an academic skill.* It is learning how to stand alone, unpopular and sometimes reviled, and how to make common cause with those others identified as outside the structures in order to define and seek a world in which we can all flourish. It is learning how to take our differences and make them strengths" (112). Lorde sets the homogeneity of the master's situated, white maleness in opposition to an unheimlich, diversified womanhood.

However, Lorde does not (and likely cannot without losing credibility) directly state that she intends for the women outside the patriarchal norm to be equated with the position of the master. This is why Lorde does not present a metaphor that posits a diverse womanhood as target and the master as source. Yet by arguing for the inclusion of the broadest range of womanhood possible, the metaphor works to usurp the position of the master. Moreover, Lorde's metaphor necessarily involves displacement rather than a mere reclamation of natural rights, in part, because of the dissonance of the site of its launch. Although Lorde asserts her inalienable claim to her words and to her subjectivity as tools, she does so from within the metaphorical "master's house." Thus the shape of the metaphor set is inherently constrained by this moment.

To delineate the shape of a set of points in physics, a "moment" is usually defined in relationship to a fixed point. The center of the moment is identified as the actual point about which the force causes rotation. It is also understood to be an axis about which a force may cause rotation. From this perspective we can imagine diagraming the function of a metaphor operating similarly to Holloway's description of story when drawn around a cylinder, except the conveyance happens in an instant rather than occurring over time. Thus the arrow representing the movement of characteristics about category A across to category B appears as a line drawn around the surface of a cylinder, with A being at the beginning and B at the end and with A and B represented side by side. The metaphor seeks to convey attributes of A—the master's situated, white maleness—by calling them into focus. Once the set of characteristics is in focus, metaphor uses tension to generate enough energy to snap the source set onto the target. As a result, Lorde's metaphor—the master's house = academy; the master's tools = traditional academic discourse—conveys the master's situated, white maleness onto the academy and its traditional discursive operations.

The logic of the metaphor is as follows: *If* one cannot use the master's tools to dismantle the master's house *and* the master's house must be dismantled, *then* one needs a different set of tools to dismantle the master's house. *And if* the master's house is the academy *and* the master's tools are traditional academic discourses, *then* one cannot use traditional academic discourses within the academy if one hopes to transfer its benefits to another. The benefits suggested by the metaphor are tied to inclusion. Lorde describes an embodied presence, which though largely absent from the conference, she herself represents. She states at the opening:

> I agreed to take part in a New York University Institute for the Humanities conference a year ago, with the understanding that I would be commenting upon papers dealing with the role of difference within the lives of american [*sic*] women: difference of race, sexuality, class, and age. The absence of these considerations weakens any feminist discussion of the personal and the political. (110)

Her embodied presence is challenged rhetorically (and otherwise) by the construct of the master. And it is this point about which the metaphorical force causes rotation. Lorde is one of two black women presenters at a conference on the only panel at the conference "where the input of Black feminists and lesbians is represented" (110). Thus, she establishes the ground for her metaphor with a critique of the practices that lead to the exclusion of diverse voices in such settings. The entrenched cultural distinctions between category A—the master—as source and the diverse womanhood for which Lorde purports to speak from within hallowed academic halls generate enough tension that Lorde, as the embodied presence using her discursive practice, rotates around the central axis of the metaphor to appear beside the source. While the metaphor is constructed as we have said, by creating an equation that links the master's house and his tools with the academy and its traditional discursive practices, Lorde's speech act itself has metaphorical import. Consequently, in addition to the set of attributes that the metaphor calls into focus and conveys from the source to the target, the site operating alongside the articulation of the metaphor work to displace (rather than dismantle) the master in his house.

What then are the takeaways from this consideration on metaphor? Traditionally, we have been taught to recognize metaphor as a precise literary device, an ornamental embellishment that refines a piece of writing. The fact is metaphor transcends our use of language. Metaphor is both organic and constructed. It appears at the beginning, before language, before image, before articulated thought. Metaphor emerges at the cognitive level and generates connections between the disorderliness of human experience. In other words, at its base, human cognition is metaphorical. People comprehend things by understanding what something is like. This act of correlating previously disconnected things is the way by which people encode new information. In essence, the mind draws a line from what is familiar to that which is unfamiliar. This function of a line is one of the primary concerns of a writer. The line, then, creates a set which begins to categorize things that might otherwise appear as random.

Moreover, metaphor offers writers a tool for engaging precognitive matter(s) and that capacity is increased as they become more adroit at utilizing metaphor. Writers are masters of metaphor. But, for the most part, writers are oblivious to the fact that they operate at the root of human cognition. Because they are largely unaware of this reality, writers yield this cognitive terrain to scientists and psychologists and advertisers and the like. Yet a writer who is able to manipulate the form of metaphor is able to delve more completely into human cognition than are writers who treat metaphor more esoterically. The writer's work is in capturing the energy found in the elastic potential of language to move beyond the cerebral to more fully engage the human experience. Understanding the mechanics of metaphor helps writers convey meaning more effectively and efficiently. If writers exploit this simple point, their writing will appear natural because it is doing the hard work of the brain to demonstrate the linkages between what a reader already knows and some new thing.

The degree to which one creates effective metaphors helps to determine the depth to which a reader can engage the message. Along with notions that get explicitly articulated as text, metaphor carries a vast array of connotations, so metaphorical thinking can begin to function more telepathically, which is to suggest communication that operates beyond the limits of words. This organic energy can be harnessed by a well-constructed metaphor. Metaphors are engineered like bridges built by explorers seeking pathways to a great unknown.

By contrast, story is employed as a tool within society, speaking to people across cultures and across time. While on the one hand, metaphor reveals in a moment often surprising links between disparate things. We might consider poetic uses of metaphor that reveal the truth of human experience without attempting to assert itself as somehow complete. On the other hand, story unfolds over time as a justification for the fairly capricious and arbitrary representations used to fashion a narrative arc. Story seeks to reconcile experiences by placing them into a construct that seems ready and able to house them. While story may employ metaphors and story itself operates metaphorically, metaphor is apart from story. Metaphor is simultaneously larger and smaller than story. Metaphor is cognitive DNA—conceptual building blocks—from which meaning is drawn out of experience. The challenge for contemporary writers interested in the art of storytelling is to develop systematic approaches that harness the vast creative capacity of metaphor.

12

The Glittering World of Synapses

Lyncia Begay

The soils offer the nothingness that droughts often yield, causing plants to pull into themselves until they invert, becoming nothing. I can see this nothingness in the miniscule flakes of skin that run up my arms and flake from my body as they crumb into dust.

Running along a dirt trail that peters into ponderosa forests, I phonetically emit an inventory of plants that I carefully pronounce in Diné[1] bizaad.[2] I speak so they can hear their own names:

waa'[3]
nidishchii[4]
k'įį'[5]
chiiłchin[6]

Precious is our recall. The names of nanisé,[7] land formations, precious stones, and all within Diné bikeyah[8] is to honor the fractal atoms that perform the geometry seen in lightening, synapses, and the dried riverbeds that lie across the southwest to perform systems of response.

In my own reply, I am following a trail that runs along a river that's older than this city. At my current locale, from memory these waters haven't reached this juncture since I was a child. The Rio de Flag, a colonial placeholder for the river I am following, was once not only a river but a common floodplain yielding millions of gallons in the years before water abstraction became common in Arizona. It's now a phantom ditch, collecting the petrochemical runoff from the city's streets while at once treated as a repository for effluent waters that's toxicity cannot be reclaimed. Water that is reclaimed is often used by the city of Flagstaff to sustain large stretches of grass and to create an artificial landscape of snow for a contentious ski resort named Snowbowl.[9]

Since the onset of settler's arrival, this area's biomass has been radically transformed—subverting not only the natural ecology but the trajectory of Dinétah.[10] Beginning in the 1950s, a small dam was built diverting the stream further into itself, so it wouldn't spread into the newly settled suburban neighborhoods. The margins of these redirected river systems are a reminder of how rivers today parallel an infrastructure that is inorganic often rendering the natural systems of response that were once in place.

One such system includes the nearby Little Colorado River an offshoot of the Colorado River,[11] now intermittent.[12] In the Diné universe the Colorado River is a necessary vein that eventually flows into smaller channels, comparable to the body's venules.

As I run alongside the diverted Rio de Flag, I can't help but think of the Little Colorado River. The Little Colorado is the in vivo venule carrying water to the various plants roots. From here the water undergoes photosynthesis as the leaves of plants present small opening called the stoma that dilate to absorb necessary carbon as larger quantities of oxygen and water escape from the same opening to enter the atmosphere. This vascular transport of water creates transpiration that flows back through a vascular system comprised of cells that are indivisible, held together by the same mystery that asks what the metaphoric heart and even brain would be within this interconnected system of cellular nations. From my flesh a series of stories respond.

DNA—those smallest living units of self, cell by cell, scattered and gathered, constellate into our bodily being. Each cell honors the negative space that allows us to perform materiality and the barriers that distinguish what our nucleus contains.

In Latin, nucleus is short for kernel. A kernel is a shell that will break beneath sodden baked soils, working to reveal the constellating constrain to root, bud, bloom, pollinate, and repeat. A nucleus containing DNA instructs beings into accumulating matter until they are materially whole. The nature of such instruction has always been formative to our own origin as Dine people.[13] I've learned existence often exacts what comes from a kernel, a nucleus—a sentience that is mutable and expressive among a lively cosmos. Once understood, a series of cellular nations begin to collapse into the variant reflections of protein particles who do best when the shells of appearance dissolve and there are the offerings made to the river that I am running along. These offerings extend to the deer and plants traveling this same path that I yield myself to.

That's not to say yielding and dissolving are such a simple reality. Distinction, in terms of space, involves imparting an awareness, a regard for the fluxes of the natural knowns. The wild tobacco demonstrates this well. It's a plant able to pattern its pollination during either night or day to control the kind of pollinators it receives. Wild tobacco also uses its nectarine to organize the rate at which hummingbirds and moths pollinate at each flower by repelling or attracting through a constituent of nicotine. In terms of mathematics, the wild tobacco's nectarine allows it to cultivate the likelihood of pollinating with unrelated plants—further highlighting the awareness that the wild tobacco's kernel holds, an awareness that seeks to diversify in terms of genetics, an awareness that acknowledges a conception of self, kin, and what is separate and shares a likeness.

This unique nucleus demonstrates something fundamentally Diné as our clan systems rely on permutations of identifying the clans of our mother; our father's mother; our maternal grandfather's mother; and our paternal grandfather's mother. Knowing one's clans is used to understand who we are and what our roles are in relationship to the earth's processes. Clans are also used to understand how we relate to each other, to find kin among strangers, and also to avert genetic mutation. While the wild tobacco's nectarine is a result of cellular response, our clan systems are derived from our memory. Memory is a reflection of the neuronal pathways that are continually tread, until they are stored, becoming a system of stories, allowed their pattern in which they are performed in the actual world.

Such varying cells are held together by a common thread, the water. Water is a composite nation that forms from rain, snow, sleet, and morning dew. While water is a changeable entity, as in a story, its outcomes are also mutable—true as the traced narratives that have created a language of nouns; of things, rather than beings. I can see how this language of nouns has stripped the rivers, the bloodstream that enlivens responses, and the ability to perform recall and the beings that respond. The rivers I see are living in an interconnected entangled physical reality that lives and dies based on who is observing and transforming the water.

As if in response, a metaphor performs along the curve of a pond as I run. Neon green is the Rio de Flag repository, like something you'd see in the Simpson's irradiated rivers. The toxicity of this city has penetrated the circulatory system of memory to perform what I imagine is bioaccumulation.

The blood that circulates throughout the whole body allows oxygen, movement, nourishment, and the experience of whatever we ingest, whereas the physical epicenter of the sensory mind, the brains as a nucleus, pries into existence like a kernel seed steeped in darkness, climbing for cohesion, for latitude. This invisible trajectory of thought is unseen whereas the brain is populated with pericyte cells that wrap around the endothelial cells that line the inside of the capillaries creating an extracellular dispatch. Pericyte-endothelial cells are tasked to protect the neurovascular system through a blood-brain barrier (BBB), a permeability that either allows or inhibits, maintaining the integrity of blood in the brain—a maintenance that retains neuronal responses, so the mind continues en masse.

This is the story of all stories as the neurons readily mirror a Glittering World[14] complex that requires memory to maintain stories. Neurotoxins, trauma, and even heat can damage these cells—leading to a breakdown in the BBB; contributing to cellular waste which eventuates in the breakdown of endothelium resulting in neurodegenerative diseases like Alzheimer's that affect neuronal activity.

Globally, pharmaceutical waste and toxins have essentially penetrated the BBB of plant systems and the people either noticing or not noticing the toxicity as part of our lives. The major veins that have branched along the earth have pulled into themselves, their flow regressing—an intermittent river that blanches, causes an ecology of plants to pull into themselves, causing the horses to macabrely drag through the arid summers along dry riverbeds, turning on their bellies, becoming ribs, and finally dust.

Along these thoughts, as I conjure the ribs of horses, the regression of fear stymies the growth of neuron cells. As I run, I recall the riverbeds I've crossed as I walked along highway 89, en route to Bodaway Gap from Flagstaff (some seventy-six miles north of my present locale). Along this way, I had walked, admissioning that the horses are the prayer. The fear collects in me as today the horses turn up along dried riverbeds in repose, in torpor, ceasing to live. I see this even as I run along a river that's often filled with the petrochemicals that also infiltrate water systems, infiltrating the vascular and this memory.

I rerun a memory from the walk[15] as I pull into the nothingness of myself and find myself separate from walkers as we course beneath a dry culvert with cement walls. It's hot in my thin cotton blouse whereas my saanii[16] skirt holds all the blackholes I've sealed with needle and thread. I continue into myself as I stare at the walls covered in graffiti, studying the notes of those who've traveled this same route. I am there sitting beside Casey and Harrison, but not really. I am in another dimension, as Faylene begins to bicker about the 3 percent that Harrison and I have used to buy art supplies from the GoFundMe Campaign. I am instead giving myself to the writing that is on the wall. The minds that are from a separate time have written words like "bitch," "fag," a heart and a few obscure tags that I receive with an impersonal affect while ignoring Faylene and now Danielle who is coupling in her own critiques. I walk toward our hoard of gear and pull out the art bulk. I begin to draw a cat in a hot air balloon, knowing it will disappoint the group-think that's affronted by the lack of anger and paranoia as I chalk into the walls. It's a dark place they want me to exist in. To draw a series of stars, a darkness that is clear and pure in comparison to what the walk is becoming is a stoic's revolt.

In place of riverbeds, there are the empty liquor glass bottles in the tunnel culverts. The green and clear shards are eerily dusty. I continue to take inventory as we nibble on our sandwiches in silence among a glittering perplex that prevents us from seeing each other. As the heat pulls just above, noon disperses, and we emerge from the culvert to follow an old highway, a dirt trail that follows the power lines coursed for Phoenix, Tucson, and California.

The clouds seem to climb pressing humidity into a stillness that encircles walkers. The sky turns grey as sprinkling causes the power lines to hiss with static. The electric power is immense. As I walk beneath its extra high voltage, I feel nauseated by the way it reaches into my bones to echo. No sooner have we reached the shoulder of the highway then a torrent of monsoon leads to what feels like a waterfall that we've stepped into as the wind causes our skirts and hats to ripple from our walking frames. The rain strips me of dust, as I remember with mirth and laughter the heavy canteen of water that Harrison had ruthlessly dumped onto the tadpoles—suddenly I am remembering who I am.

A monsoon's tangle of lightening
mirroring the inside of my heard—
as ions pulse through the mind
neurons that conduct
that perform a tapestry
and allow me to thread at the holes in my skirts
a movement response
akin to healing as the skirt reminds me of "amasani"[17]
who trailed the earth with her saanii skirt
she died from the cancer
the uranium that passed through the rivers she drank and cooked from
the rivers that branch into each other
the rivers I talk to, as they're beings
responding to plants, horses, and the prayers
that are my movement
movement of a water clan being
a Biihbitoodnii[18]
a Biihbitoodnii watching the
EHV power lines loom
while just above

thunder signatures
ask us to realize that real power
causes us to walk pass destruction and bickering
into states of vulnerability
a kind of hope that massages the broken liquor bottles into sand
becoming settled sediment running adjacent
to the fluidity of glittering waterways
that transmogrify the earth's breathing patterns
into a series of coordinated seasons
as trees oxygenate
patterning wind
creating climate
as birds migrate
becoming a prayer
a sacred speech
breathing to emit words
that pry and rise
from my mind

a healing that begins from my forefinger and thumb, poised as if I'm holding a threaded needle as I spin a pinch of yellow corn pollen into expressed gratitude, hope and empathy for a water system that I live through the same pattern of abuse and dehumanization that Diné people struggle with. Yet, mother-like is the snowmelt from the Rocky Mts., the Colorado River, and the generating monsoons that relentlessly give us the water to bathe and nourish our being. Along her rivers, I imagine clouds flocking into a thunderstorm that causes bloodstreams once numb, to tingle with the return of feeling. It's the same coordination I wish for my mother as she mirrors the powerlessness I feel as we mirror the units of cells, always wanting our whole selves back.

13

A Sense of Oneness with Sun and Stone

Leila Philip

Vladimir Nabokov loved butterflies. He loved watching them flutter down onto plants, close their impossibly fragile wings, and feed; he loved watching them angle through the evening light. He loved their heroic migrations that can span hundreds or thousands of miles. He loved watching them transform from ugly caterpillars into winged insects whose colors and patterns are so intricate and beautiful they boggle the mind. But most of all he loved the ability of butterflies and moths to mimic and trick; the way they can fool predators by resembling leaves or displaying dark spots on their wings that might be huge eyes; the way certain moths, which start out looking like bird droppings in their infancy, later change their camouflage, developing appendages that look like lobster claws; the way some butterflies even perform theater, imitating the oozing of poison through bubble-like macules, or the appearance of being covered in yellow goo as if they had already been eaten and thrown back up, their performance signaling "do not eat me, I will make you sick!"[1]

In his memoir *Speak Memory*, published serially by *The New Yorker* in the 1940s, Nabokov devoted an entire chapter to his love of butterflies, including the following passage, which established not only his adoration of butterfly mimicry but also the ways he connected it to his literary aesthetic.

> The mysteries of mimicry had a special attraction for me. Its phenomena showed artistic perfection usually associated with man-wrought things.[2]

Nabokov was aware of the technical usage of the scientific term "mimicry," which means protective adaptation as object resemblance, but for him, the trickery of butterflies and moths demonstrated something deeper than natural selection; it was evidence of a profound connection between art and science. He continues:

> "Natural selection" in the Darwinian sense could not explain the miraculous coincidence of imitative aspect and imitative behavior. . . . I discovered in nature the nonutilitarian delights that I sought in art.[3]

For Nabokov, in other words, science and art shared recognizable qualities of beauty and wonder. A butterfly's survival costumes are too elaborate to have been developed just for self-protection. In this creature Nabokov saw nature's confirmation for his own literary aesthetic—narratives that relied on complexity and trickery. In a Nabokovian flourish, he concludes:

> Both were a form of magic, both were a game of intricate enchantment and deception.[4]

While the earlier American writer-scientist, Henry David Thoreau would find the universe in a maple leaf, for Vladimir Nabokov, all the wonder (and trickery) of the world lay in a butterfly's wing.

Nabokov's particular passion and life's work was a widespread group of small butterflies known as *Blues.* Part of the tribe *Polyommatini,* Blues are found on every continent, but their name is misleading for they come in many colors including white, brown, and grey. While Nabokov's tremendous literary output and influence as a lecturer and teacher is well known from his years teaching at Wellesley College then at Cornell, the extent of his work as a lepidopterist—one who studies butterflies and moths—is only now becoming better known. From 1941 to 1948 Nabokov actually held a position at the Museum of Comparative Zoology at Harvard. In 1965, seven years after the publication of *Lolita* had vaulted Nabokov to international literary stardom, it was significantly an image of him in pursuit of a butterfly, net in hand, taken by Philippe Halsman that he wanted *The Saturday Evening Post* to run with an article about him. Repeatedly in interviews Nabokov stated that he had not one but three careers—writing, teaching, and lepidopterology.

By the end of his life, in addition to his seventeen books, his plays, stories, and poems, his movie scripts, his literary criticism, his many translations and his extraordinary lectures on literature, Nabokov also published twenty-two scientific articles on butterflies. While he never completed the two books on butterflies, which he mapped out when he was still alive, several species of Latin American Blues are still named after him. Before his work, very little was known about the life cycles of these butterflies. As the fascinating book by Kurt Johnson and Steve Coates titled *Nabokov's Blues: The Scientific Odyssey of a Literary Genius* details, in many ways, Nabokov was as much the Leonardo archetype of the artist-scientist as certain popular accounts made him out to be. Although he was self-trained, he joins Humboldt and Darwin as one of the only three scientists to have their names permanently associated with their discoveries: think Darwin's finches, Humboldt's currents and Nabokov's blues.[5]

Repeatedly throughout his career, Nabokov would make statements that complicate the usual division of science as rational and logical and art as intuitive and creative, as he did in an interview in 1962.

> In a work of art there is a kind of merging between the two things, between the precision of poetry and the excitement of pure science.[6]

Or put another way, he might answer:

> There is no science without fancy, no art without facts.[7]

What was this famous literary trickster trying to say? When and how do the questions of literature and the questions of science overlap? How might we make use of these questions in the writing of nonfiction today?

That the study of butterflies gave Nabokov a great deal of material and vocabulary within which to develop literary ideas, themes, and working ideas has been well documented. In his novel *The Gift,* the narrator is like Nabokov himself, a writer and lepidopterist. One can argue (and it has been) that one of his main themes—concern for human beings twisted by obsession—was fueled by his emotional understanding of the passion of the collector. That Nabokov would use insects underscores the fearless nature of his work. For the most part, when insects or entomological themes are used in literature they are used to convey lust or to depict things that are unsettling. One need only think of Kafka's famous character, Gregor, who wakes up as a cockroach, to see this. By contrast, for Nabokov, the butterfly was always a source of vitality and life. It is no surprise, then, that when he sat down to chronicle his life, in the memoir *Speak Memory*, the story of the birth of the writer is given almost equal weight as the story of the birth of the lepidopterist. In chapter six, he writes:

> From the age of seven, everything I felt in connection with a rectangle of framed sunlight was dominated by a single passion.

> If my first glance of the morning was for the sun, my first thought was for the butterflies it would engender.[8]

The study of butterflies did not just give Nabokov information about butterflies and moths; years of butterfly hunting and research trained him in the methods of taxonomy—the observation of physical details which a scientist studies in order to fit those details into a larger picture. This methodology supported, if not informed, Nabokov's methods as a writer, which always exalted the specific over the general; every one of his books whether in the genre of fiction or nonfiction was crafted from an accretion of carefully observed details. He was insistent upon accuracy and the historical coherence of elements that went into his fiction. As he would repeat in various contexts, "in high art and pure science, detail is everything."[9]

Nabokov was well aware of the mutable nature of empirical facts in terms of how we understand them given that science is constantly changing what we know of the empirical world. We need only consider the growing fields of neuroscience and cognitive science to see that this is true. In recent years, neuroaesthetics has begun to study the impact of art, music, and performance on brain development and evolution. Once you begin to acknowledge the possibilities of aesthetics as a player in evolution, many adaptations earn a second look. Consider the male peacock's enormous tail. While that behemoth of a tail radiating dark eyes seems excellent at attracting females, its great size and weight puts males at risk; they can hardly fly with such a tail, making them more prone to predation. Why would attraction outweigh the ability to flee? Similarly, in humans, the square jaw of males is considered a universal sign of attraction increasing the probability of the spreading of that particular male's DNA. However, square jaws, which are associated with higher levels of testosterone, also indicate a weaker immune system, meaning the male in question is more at risk. In many ways, Nabokov was ahead of his time in re-considering Darwin's straightforward theory of survival of the fittest.

Much of Nabokov's writing models the benefits to a writer of staying alert to science. We live in a moment when our understanding of how the brain makes and records memories, experiences time, and takes in the world has dramatically enlarged narrative possibilities for writers of nonfiction in their approach to writing about everyday reality. What Nabokov reminds us is that when writers fully engage the complexity of the real, they don't have to sacrifice narrative trickery in order to reap artistic gain.

Now, as Nabokov himself was fond of saying, dear Reader, let me intrude. I have my own brief butterfly story to tell.

* * *

It was one of those late July mornings in New England when the world is so steamy and green the garden seemed to be growing before my eyes. The cats were lounging in the rhubarb, the dog was stretched out in the shade, and I was standing, coffee in hand, admiring my tomatoes. They were already taller than me and studded with green fruit, some the size of my hand. Then I noticed that the top of one of the plants looked strangely bare. I walked over to inspect, puzzled, and yes the upper section of the plant was not just bare . . . it was missing! Deer? I wondered. A rabbit could not reach that high, nor could a woodchuck.

Then I saw it, a long green caterpillar the width of my finger. I looked in horror, it was the perfect green of the tomato plant, camouflage extraordinaire, but with a series of white stripes and white-ringed black eyes along the back. At one end, it sported what looked like a long reddish horn. No bird would tangle with that bug, no way; it looked as fierce as a samurai. I almost jumped out of my skin when

I realized that it was not alone, similar caterpillars of similar size or larger were stationed throughout the entire plant. I knew I had to take action. Taking a deep breath to quell my rising sense of panic, I began to pull them off, which took a surprising amount of effort; each was latched to the plant with two rows of tiny claws.

As soon as I had pried them loose, I whipped out my phone to get some pictures. They were so horribly beautiful with their spotted sides and that long slightly curved red horn. I made a note to look up information about them. But first, I had to save my tomatoes. I threw the thick caterpillars onto the middle of the driveway where I figured birds would get them. Or they would shrivel in the sun. The cats, interested now, poked at them a bit, but quickly pulled their paws away. The dog, annoyed at being disturbed from her nap, dutifully lay down on the side of the drive to guard them. Within an hour when I went back out to check, the dog was asleep and the caterpillars were gone. Eaten I thought, with satisfaction; if they were going to murder my tomatoes, I would be equally ruthless.

Later, I texted the picture to a dear friend who prides himself on his extensive knowledge of all kinds of matters. "Swallowtail butterflies," he texted back. As soon as I read the words I was devastated. Here I had found the caterpillars of butterflies that like the monarch were struggling for survival and what had I done? In my mammal greed, I'd rushed to kill them. I felt sick. Why hadn't I saved them in a jar and let them grow into beautiful butterflies? For the rest of the day I was mortified at how I'd been so quick to anger, so protective of my garden, so rash.

That evening when I went out to water I found more caterpillars, and more tomato plants completely denuded at the top. Ten caterpillars in all, this group as thick as small cigars. Their fat green bodies still made me shudder, but this time I carefully pulled them off and placed them in jars with holes punched in the top for air. I fed them tomato leaves, which they appeared to love and water by soaking a piece of cardboard and laying it across the bottom. I was determined to let them grow and become chrysalises, then hatch and fly away as beautiful rare swallowtail butterflies. I worked hard to garden with the ecosystem around me, not like some suburban fool destroying the environment with miracle-gro and pesticides.

That night, after inspecting my jars of growing caterpillars who appeared to be eating happily, I texted photos of them to another friend, who was a professional landscape gardener. Her text was immediate, “YUCK. Tomato Hornworms KILL immediately!!!”

That was when I did the smart thing that my college-aged son would have done immediately, and checked the internet. Yes indeed, I had tomato hornworms that I learned could consume an entire garden in days. Throughout the week I checked the garden three times a day and pulled off so many I stopped counting. This time they went out on the road, food for the birds.

And what, you may ask is the morale of this story?

I loved the idea that these caterpillars were swallowtail butterflies and I prided myself on gardening with nature not against it and that was the story I was building in my head.

As many have documented, the noted biologist and writer E. O. Wilson called humans a storytelling species. Our brain longs for narrative for we demonstrate a deeply ingrained instinct to seek pattern. Throughout our evolution, story helped humans organize and convey all kinds of information quickly and efficiently; things conveyed through story are memorable because listening to story is pleasurable. Stories convey survival information such as whether to hide or run when that big catlike animal with stripes prowls into view. But we are social animals, and stories convey information useful for living with other humans, information about history, culture, tribe, family, and the mystery of character. On a personal level, we make sense of the world through story. Beyond giving us a survival advantage, we think in story because it enables us to make sense of the hard things about being human—that time spares no one and everything we most love, we know we will eventually lose.

My story of the swallowtail butterflies that were actually tomato hornworms reveals both my romantic ideas about nature and my fear of environmental degradation. Maybe I can't stop climate change, but I can save endangered butterflies. A small act, but something, or so I was telling myself. The problem was I had rushed to conclusions and actions without the facts and I'd been led into a story that was not just so limited as to be foolish, but was harmful. If I had taken the time to investigate with the precision Nabokov spoke of, I would have discovered that swallowtail caterpillars look nothing like the fat green cigars chomping leaves in my jar. I could have googled what eats tomato plants and seen a picture of a tomato hornworm pop up along with the warning in capital letters—GET RID OF IMMEDIATELY.

The point is not just that a story grounded in the wrong facts can have bad consequences, it is that by not taking the time to attend to the real, I was losing out on what was important about the experience and what could led to a more complex and interesting story in the end.

What would Nabokov make of my tomato hornworm tale? First, he'd probably roar with laughter. Human folly gave him a great deal of material after all. But more seriously, I think he would go back to his oft-quoted stance that the writer would do well to engage with science. As far back as 1952, he was making the case for the overlap of science and art.

> Does there not exist a high ridge where the mountainside of "scientific" knowledge joins the opposite slope of "artistic" imagination?[10]

Vladimir Nabokov asked this question in his *New York Times Book Review* on December 28, 1952, in which he considers a book about the great American ornithologist and artist, John James Audubon. Audubon interested Nabokov because as a nineteenth-century naturalist and artist, he bridged science and art, the vocations of naturalist and painter. The point Nabokov is pressing through this rhetorical question is in part answered by even a cursory look at Audubon's two most famous books, The *Birds of North America* and the *Quadrupeds of North America*. To produce these books, Audubon followed the practices of a nineteenth-century naturalist, trapping, hunting, and snaring birds and mammals in order to study them and paint them accurately. Yet unlike his British predecessor Mark Catesby, Audubon did more than depict his birds as specimens. In Audubon's paintings, birds playfully preen and swoop, eat and coyly peer at one another, perch, or attack with ruthless accuracy. Animals are painted with similar variety. Although in the *Quadrupeds*, almost all of the mammals are depicted with open mouths, not because Audubon felt it was aesthetically pleasing to draw them baring their teeth, but in order to include information about the size of their jaws and the number and type of teeth. The books, which even then were tremendously expensive, were bought and displayed by the wealthy in nineteenth-century America, as much to instruct viewers about facts of nature as much as to delight them with works of art.

Audubon, the artist-naturalist, also interested Nabokov because clearly he saw in this early American artist-naturalist a version of himself—the scientist-artist—and more to the point a set of questions that interested him. Namely, what does the intersection of science and art look like? When and where do the questions of literature and the questions of science intersect? Of what use is the study of science and the scientific method to the writer? These questions are of value for any genre of writing, but they seem particularly important for nonfiction today, an area of writing plagued by a name that offers only negation, completely disregarding what is literary about the genre.

And where does this leave us? I hope I have shown how Vladimir Nabokov, that literary giant so fond of narrative complexity, who saw all forms of art as elaborate deception, paradoxically spent much of his working life in a devoted study of the real. At the end of the day, like the poet-naturalist, Henry David Thoreau, who measured the temperature of the pond each morning, Nabokov was a student of nature and a relentless observer. For him the value of science to the writer was not just the discovery of material, it was the ways in which it taught the writer and the artist a methodology of close observation and attention—a message, which I think, should wake up nonfiction writers of all stripes. Because to privilege what can be invented over what can be discovered in an exploration of the world around us is to ultimately limit one's imagination.

Nabokov was a master of enigma, and he loved paradox, but when he wrote about butterflies, his writing was uncharacteristically direct, even Emersonian in its sense of wonder and awe. It seems fitting then to take a pause from further interrogation and to close by savoring his own words from *Speak Memory* where significantly, he closes chapter six with the following description of butterfly hunting.

> And at the highest enjoyment of timelessness—in a landscape selected at random—is when I stand among rare butterflies and their food plants. This is ecstasy and beyond the ecstasy is something else, which is hard to explain. It is like a momentary vacuum into which rushes all that I love. A sense of oneness with sun and stone.[11]

14

A Gardener's Education (Animal Body)

Marco Wilkinson

To write the whole body at once is the monumental effort of an anxious animal. The animal in me is desperate to tell this story whole, wants to find its limit and rebound inward, revise, perfect itself into an eidolon of itself. But the plant in me only ever essays, probing at this memory and that, looking for sunlight in the canopy and sustenance in the shadows, branching ever branching, extending past itself. Trying can be trying, but it's all that ever is.

I was an exceedingly anxious old person in my twenties, obsessively dwelling on what had already passed me by and on my mammoth responsibility to a future I felt unequipped to confront. At twenty-one, I graduated college with an English degree, a crippling depression, and a glimmer of hope in the form of an organic farm internship doused by the horror and disapproval of an immigrant mother who had done everything in her life to escape the miserable farm of her childhood. At twenty-four, I found myself floundering back at home, finding fleeting contentment in exhausting myself hiking in the woods only to contract Lyme disease (and with it meningitis and Bell's palsy), and walking across Spain on the Camino de Santiago glorious like a god walking the newborn earth only to return home and crumble back into a heap of neurosis. At twenty-six, I was already old and resigned to having lost everything, depressed and ever caught in regenerating that depression in a body and mind unable to change. At twenty-six, I decided to garden.

I could not imagine this fragile queer depressed body living past forty and so I died young, accidentally, tragically, quietly, unnoticed a million times in my head. This story-body perished and revived into just the same form over and over. Now, just having turned forty-two, my matter-body's unexpected continued presence forces the matter, and the story-body of my future propels forward, generating new forms and fantasies. "I can't imagine being eighty." But I do, and this time-body grows forward relentlessly from one story to another. Every form the final one, a corpse, until circumstance that would pin us forces growth in a new direction, like water finding some furtherance, fountain or well, defined by a cascade of little failures.

Grow or die: this is the first lesson I learned on my first day training as a horticulturist in my first botany lecture at Brooklyn Botanic Garden, taught by Barbara and Wayne, who over the ensuing weeks will bring into the classroom a mango tree on wheels brushing the ceilings of the basement classroom, a potato sequestered in a dark cabinet for a year resulting in a ghostly architecture of searching stems, vines and creepers and clamberers all conniving for light, and more. We learn about vascular systems and photosynthesis and genetics and taxonomy, but ultimately we keep returning to this first fundamental point, that while animals can move, plants can't. Or at least not in the same way, because while animals grow and develop into mature bodies that, though changing, are fundamentally regenerating and renewing themselves, plants grow ceaselessly, their forms only ever provisional ghosts of what we think a mature oak tree, say, will grow into. We talk about the "habits" of trees, that is, their overall shape, but a habit is just that, a tendency. The actual body of an oak tree is hydra-headed, endlessly generating upward and outward, a new body every season, death being not so much the wearing out of the old machinery but rather the collapse of inevitable overextension, the limb that reaches too far out and eventually rends open trunk down to the root.

In a seed there are two poles, like north and south of a world. They define the limits of extension. Extending they generate a magnetic field of force around them. Generating this field they gather affinities. Gathering is another word for chemistry, the chemistry they develop with the world around them. Every seed, from pea to pin oak, from Sempervivum to Sempervirens, has from the moment of germination one axiom to follow that overrides all others: Grow or die.

These two cells are the root and stem apical meristems. Crack a bean open by slitting two fingernails into it lengthwise and inside you'll find a plant in miniature. Folded like praying hands are this future plant's first two leaves, and hidden at the heart of this prayer is a cell. Pointing like a little finger is the radicle, root of all roots, and at the tip of this finger gesturing toward the future is a cell. Both of these cells are meristematic, meaning as one might surmise from the "stem" planted inside it, that these cells have the ability to transform into every other kind of cell a plant might need for itself to be itself. Potentially, each of these cells might make a leaf or a stem or a root or a pistil or an anther. But this energy once unleashed is purely kinetic. It must move relentlessly forward.

Activated by heat and water, a seed germinates. Enzymes break down stored starch or fat into sugar to fuel these two meristematic cells, which busy themselves dividing into more meristematic cells like bubbles in the churning wake of a boat, each of which in turn transforms into specialized interacting parts—stem, leaf, and root. The stem apical meristem clasped in those praying leaves launches heavenward; the root apical meristem, fingertip of the future, drills down to the center of everything. From that moment there is only one motion—forward—and one imperative—generation.

Do plants know time? Somehow I imagine that their time is space, the future merely a function of farther. The pine candles lengthen into the spring and needles emerge from their papery casings in bundles of two, three, four, or five into open air. Sugars coagulate in photosynthetic reactors and flow backward as if the work of the present was to feed the past. Inward and downward sugar flows, down to the roots and some 50 percent of all sugars are given away by plants to the bacteria and fungi crowding in the rhizosphere, the root-world. David Abram recounts that Balinese "spirits" are not incorporeal but knitted into the soil and life of the earth, the very ants that are fed with ritual rice offerings outside of homes. Past is not a time left behind but a space to be nourished.

At Brooklyn Botanic Garden: Rob taught me to paint mulch with a pitchfork in swinging arcs over flowering quinces. Nancy taught me that imperialism is hidden in plain sight in plant names. Ann taught me the strength of compassion, doing for me what needed to be done when I accidentally killed a baby rabbit who jumped in front of my lawnmower. Erik taught me to identify wood asters by their zigzagging stems. The Rooster taught me that to be a fool is the most blessed thing on earth. Mike taught me that being queer and using power tools are not mutually exclusive.

Despite all these lessons and an anxious desire for competence and a vocation, tasked with putting my whole life together, already old in my mid-twenties and rudderless, all I managed in my first year of training in gardens and beds and landscapes was to look at mosses crowding the space between bricks in Magnolia Plaza. When I think of all my teachers that year, I think mostly of Dr. Nelson, who had been at work for several years already on moss classification when we were introduced, a young gay horticulturist-in-training and an old gay botany professor emeritus and bryologist-in-training. For decades as a botany professor at Brooklyn College, his work was on plant evolution, but here in this eighth decade as if a new seed had fallen on old ground this sprout of inquiry flourished under a microscope lens picking out opercula, archegonia, and lamellae. Every afternoon his hunched cardiganed back and pale fleshy bespectacled face shuffled in smiling to his bench in the Science department to peer down through compounding lenses, turn to a taxonomic key, return to the microscope, turn to the key, back and forth. I hovered behind him, rootless and anxious to learn what I could about growing rootless greenery and even more pleased to spend time in this man's presence. Dr. Nelson, professor of plant evolution in epochal sweeps of time, who in his eighties decided to investigate small things: mosses. I felt like a small thing too in the world, capable of only the most circumscribed orbit.

The thing about mosses is that, being small and rootless, they must make the most of every moment, every iota of space. Without vascular systems, each cell must be proximate to water and so their leaves are only a few cells thick. Without storage capacity, as water comes and goes mosses unfurl and wither pulsing in a water-wind, endlessly engorging and collapsing with the presence and the absence of water. They grow together and develop minute ridges on their leaf surfaces to hold as much water as possible but are ready to lose it all. So tiny, stepped on, blown away, grazed by minute ruminants, many mosses are totipotent. Any cell unmoored might in favorable circumstances regenerate a whole new plant. (Is it wholly new?) Perhaps newness is less a function of genetics and more one of the unfurling dialogue between being and space. There is an aspen in the West, across hundreds of acres, contiguous in parts and cut off wholly in others by development, whose life belies sameness/difference.

Stanley stood there quietly, small and bent maybe below five-feet tall, gazing out at us, his small mid-day audience, with the long gaze of nearly a century. The golf cart had ferried him out here to the Cherry Esplanade a little too early on this warm sunny September day. He stood there patiently as a flurry of staff and interns danced like ants around him, setting up mics and assembling chairs, creating an impromptu stage, an impromptu event.

We were all gathered in the shade of three titanic bodies: a colossal dying maple and its summation of the past, an oak sapling and its future ghost hovering in the air, and Stanley Kunitz and his ninety-eight years of animal life encapsulated in birdlike fragility and a sphinx-like gaze. In 1918, to celebrate the end of the Great War, the Brooklyn Botanic Garden had planted an allée of maple saplings, fresh promise of strength and stability to mark the end of bloodshed—eternal markers of the end of the war that would end all further war. By 2002, they had grown to towering heights, eighty or a hundred-feet tall, and every year saw storms shear off more and more massive limbs that had crept cell by cell upward and outward past a tipping point. For safety's sake, these watchmen for enduring peace that never came had to come down. New saplings would be planted in their place, new hopes for peace (even as the clouds were gathering) in a time of turbulence. That day, in 2002, with all of us still somewhat shocked and dazed, some of us still in mourning, Stanley—lover of trees, gardener of poets, and poet of gardeners—had come to stand his body before the first of the new century's memorial trees and the last of the old century's armistice trees (fourteen years his junior) and measure out life in three different ways.

He spoke softly in a drawn-out high voice about being a child and watching the men make their ways down the streets at night lighting gas lamps. He spoke about peace and time. He read his poem, "Snakes of September." Stanley, this little animal, spun out to his full form and for decade after impossible decade falling back in on himself like an aging star simultaneously eulogized and christened the old and the new laid out in soaring lignin and future limbs. Still, moving in words, he reached out with his vegetable self, that being that mysteriously escaped the bounds of his body, continuously growing and growing, older than the oldest tree that day, his reach extending into all the other little animals there that day, all variously imploding, including a young gardener struck by life's measure in the body of a poet under five-feet tall, in the soft warm frame of blood and nerve and concentrated time.

Though I want to write of plants and my education in them, about the way they are generative beings and how this fact has continued to astonish and puzzle me with its implications, about how space is their domain, I am an animal and I feel trapped in time, before and after and then. Can an animal write space?

I imagine myself as a plant and what my life would look like. At one end, a head, a mind: always craning up to catch the sky, the neck lengthening until two nubs appear like swollen glands on either side, those nubs growing into prongs that unfold into limbs and two sets of five fingers. And then again as the head continues its skyward journey, until there are four arms, then six, then eight, then ten.

At the other end, my genitals I suppose: lengthening and erupting with proto-plastic appendages from rippling hips that start running before newly formed toes hit the ground, from my anus a procession of buttocks mound and recede like a double-file mountain chain. I am a monster I think, whose navel lies in a hospital maternity ward in Warwick, Rhode Island, and whose winding torso runs to Providence and then New York and then the Hudson Valley and then New York again and now here in Oberlin, Ohio, where my head rises up to the sun and my latest hands stuff food into my gaping mouth. Meanwhile my legs grip tightly in the West Warwick of my childhood, the Uruguay of my parents' early lives, the stowaway boat crossing the Atlantic and the Canary Islands of my great grandfather's escape, ever backward.

While each writer may be animal in their quest for wholeness, a writer's life is vegetal, always extending past itself in a wreckage of divergences and dead ends, always with one bright green cell like a lantern pushing further and further into the unknown. The story-body probes and ventures with its curiosity into the unknown, bends under its own weight and bursts into an array of other lives, flowers and sheds seeds into the wind, all as the animal body remains curled at its desk.

After Brooklyn Botanic Garden, having not secured the open botanic garden assistant position that another intern got (and then left a year later never to continue with horticulture), after a brief stint in the plastic horticulture of hanging fake holiday decorations in the lobbies of mid-town office buildings, I went on to continue studying at Stonecrop Gardens.

Not the bloom but the thorn, not the bee but the sting: the memory-cask.

When I applied to the Stonecrop Gardens program everyone said it was known as "horticultural boot-camp." My interview lasted two days. In the middle, dinner with the director, Caroline Burgess, offered a break from the pressure: dine-in at Ming Moon, the divey Chinese takeout where just as I filled my mouth with General Tso's tofu, Burgess asked, "And what do you think that is?" She pointed to the needled tree by the waterfall in the cheap wall painting above us that I hadn't even noticed until then

First lesson: Plants are everywhere.

Caught!

"A . . . pine?"

Pursed lips.

"I would have thought a Sciadoptys," she said

I failed. But a week later I got the call that I'd been accepted.

Second lesson: Failure is a pedagogical tool. (A lesson I would learn over and over again that year.)

When I got to Stonecrop Gardens and moved into my apartment on the grounds the other interns who had been there a while called it "the convent on the hill." We worked all day in the gardens and walked through the gardens after the workday until dark with notebooks and pens at hand trying to catch each bloom firing from Caroline's lips as if we had never labored in them or set eyes on them. Caroline knew everything.

In the book of horticultural koans, this would be my entry:

She caught me on a variable spring day, in the propagation greenhouse, the kind where clouds whip through the sky like horses' tails in the sun. The benches were bursting with seedlings, and there a tibouchina tree growing out of the middle of the house, pressing its branches up against the glass roof aiming for the vents, dropped the last of its scarlet leaves that had been falling for the last few weeks as new green leaves emerged and it got ready to burst into violet bloom.

Every day one of us interns in this house spent way too much time picking individual leaves off the seedlings underneath. Caroline came up behind me as I was collecting more tibouchina leaves brought down by the sudden cool gusts that blew in through the open roof vents and made me jump when she asked sweetly,

"It's a sweater kind of day, huh?" She pointed to the cardigan I had thrown on at lunch because the day had taken a turn from the sunny to the chilly.

I smiled and commiserated good-naturedly. "It sure is, isn't it?"

"PLANTS CAN'T PUT ON SWEATERS! WHY ARE THE GREENHOUSE VENTS OPEN?" she roared.

I was shocked, then hurriedly began turning the cranks and pulling the chains to close all the vents. Fifteen years ago and that has been the most valuable lesson I have learned as a gardener: Plants can't move. Plants can't put sweaters on.

The apical meristem cell climbs a ladder of its own making as the daughters it leaves behind transform into the xylem and phloem cells of a stem's vascular system, the parenchyma and stomata of leaves. At each rung in the ladder, more meristem cells are left at the base of each leaf, waiting. As long as the apical meristem keeps staring into the sun launching up into the void, it scatters down like plumes of rocket exhaust a shower of hormones call auxins that suppress the growth of the meristem below it.

I was on a trajectory skyward. From Stonecrop I got a job gardening at the Garrison Institute, an interfaith retreat center housed in a majestic former Franciscan seminary overlooking the Hudson River. From there I became a horticulturist at The Cloisters in New York City. For four years I worked in two incomparable settings overlooking the flowing estuarine river pulsing from Albany to New York Harbor and from New York Harbor to Albany in waves of salt and sweet. In those years I learned a thirst for independence and the importance of being kind, two lessons learned through contrast with my experience. I left both places because I could not handle humans, a failure that still pushes me forward.

When the willow branch breaks in a storm and is carried from its former streamside home down the current, it may catch on stones, get tugged under to the riverbed, chewed on by a small animal, laced into a beaver dam, or it might wash up in the muck, lucky enough to be half-buried in strangeness. If so, then all the meristem cells tucked beneath thin yellow skin transform, give up plans to form leaves and shoots, and instead lay down roots in a new home. Transplantation requires transformation. Only one law: Grow or die.

My partner got a job teaching in Oberlin, Ohio, and everything changed. Landing there, I found myself working at George Jones Farm, a nonprofit educational organization, in town, where I can honestly write that I gave every bit of myself to the endeavor of growing food and growing farmers. And still . . .

Failure on a farm is a given.

Machinery broke down. Irrigation malfunctioned. Weather knew nothing of agriculture. The farmer's market got rained out. The restaurant (and its orders) went under. My co-manager suffered from ADHD and bipolar disorder and I didn't have the skills to manage my fellow manager. I was a pessimistic perfectionist and failure taught me, brutally.

Farming is a vocation of the daily surmounting of little failures. The lettuce unfurls ruffled leaf after leaf, the cucumbers swell from one end to the other, the wheat rises and bristles and goldens. It's the human that fails.

Once, when the intern arrived late to close up the chicken dome tractor, it was already dark and all the chickens had long ago roosted. No need to chase an errant hen in the sunset light this time: Easy. In the morning she came to me both upset and struck dumb. "Come see." I saw. At least ten of the twenty or so chickens were dead, splayed out on the ground, while the remaining hens were packed into a corner and obviously distressed. What we didn't see for minutes while staring was the still body of the owl over two-feet tall, in the very center. Its eyes barely registered us there or cared about our presence or even about the remaining hens. It had done its duty in the night and was satisfied with the work it had wrought. When we did realize the terrible truth at the center of this death, we cautiously opened the door as if some ferocious angel were about to emerge and cast divine judgment. The owl noted the open door, hopped once, twice, three times, spread its wings and left us behind.

If wind, hail, fungal or bacterial disease, a slug or snail's rasping mouth, animal teeth, or pruners removes the apical cell, the meristem cells below awaken, furiously initiating their own upward launch and the wake of green flesh bubbling into being behind. This is called branching, the way a singular failure occasions a multiplicity of existences.

Failure is *dandelions* rampant and unstoppable *quack grass* because human order *amaranth* is unnatural *overwintered vetch* in a *mugwort* world of teeming *nut sedge* life which only *curly dock* wants to repair what is *thistle* disturbed. The soil succeeds when *fibrous roots* anchored and *chickweed* covered, when nutrition *taproot* cycles through various horizons and little by *mycorrhizae* little relationships lead *springtails* to living tissue *nematode* settling into *bacteria* humus.

Agriculture is disturbing. Soil is undisturbed.

Failure has a way of opening sideways paths because what else is there to do but move on? In a small town, moving on is a painful public occasion. We were all laid off in November; in January, demands were made of me to produce seed orders and crop plans for a place that wasn't paying me, and in February I was offered my job back with a 25 percent pay cut and the loss of my assistant. Moving on became a little easier.

When the apex falls,

The gardener's hands fall off and what do they become? The callouses slough off into vellum folios, the clay caked under nails turns into ink, the nails now nibs crack in vertical slits, fingers blunt and sharpen and blunt and sharpen as they trace out a thousand lives true and imagined.

When the animal body fails, it is salt invaded by water. When the plant body fails, it is branches. When the time-body fails, it is reinvention into a new story-body. To write is to right oneself after a fall, but always into a new body in time. When the story-body fails there is always another story-body.

The streambanks are crowded with willows; just beyond the lightning strike, a green sprout emerges.

the submerged spirits rise.

The farm was a failure, but I wasn't. I moved on to writing and to teaching.

Failure has a way and that way is called branching according to causes and conditions. Some will go forward while others will not, some will fledge into flowers while others will not, some will bear fruit while other will not, some will grow heavy while others will not, but all will proceed ever forward.

Now I teach sustainable agriculture at Lorain County Community College. I teach soil and fungus and the joy of rot. I teach shit and regolith and peat moss. I teach scarification and stratification and mineralization. I teach wonder and improvisation and failure.

Failure teaches me to trust in the plenitude of the periphery and let go of the death in the very center.

I write and each time it is a branch testing the air, a root testing the ground where failure to fully narrate myself causes another turn, another story-body to come alive.

I teach and these seeds disperse:

Lori's indefatigable drive and cheer opens doors and purses and double-dug beds to create a community garden where children and adults learn and eat and learn to eat.

Jim is an indigenous micro-organism evangelist, telling anyone and everyone about the benefits and the joy of intentionally working with the microbes, fungus, nematodes, arthropods, and more in the soil.

Sylvia, Jessica, and Nicole give their hearts and souls and imaginations to growing food and relationships at the food pantry's community garden.

Ginette starts one market garden in her backyard, making lasagna beds in one corner, burned-timber raised beds in another, with chickens and a little greenhouse and beans drying in her garage, only to move and in mere months recreate the whole operation on her new land.

Mandy, who almost dropped out of my class, raises chickens now and sells eggs at the farmer's market. She grows sprouts and microgreens, sells cheese on the side, and delivers eggs to my door.

The whole body cannot be written.

15

The Secret Lives of Stories: Rewriting Our Personal Narratives

Frank Bures

Around the time our daughter turned four, she started making what seemed like odd requests. "Tell me about the sad parts of your life," she would say at the dinner table. Or, "Tell me about the scary parts of your life."

This phase went on for a while. I played along, telling her about my appendectomy in Africa, the time I almost fell off a cliff, the time I got a fishhook through my finger. We talked about deaths in the family, and she would sit with her eyes wide, not saying a word, listening as if her life depended on it.

It wasn't until I'd gone through a whole list of broken bones and broken hearts that I realized what she was really asking: How can I deal with sadness? What should happen when I'm afraid? She was looking for scenarios out of which to build her own. She was looking for directions about which way to turn when she reached those crossroads herself.

After thinking about this for some time, it occurred to me that I had done a similar thing. It was in college, when I discovered that I loved to write. I wondered if I could do it. I wondered, “How do you do it?”

In search of answers, like many beginners, I approached other writers and bombarded them with questions to learn their secrets and to find out how they got where they were. As it happened, there was just such a writer in the town where I went to college, south of Minneapolis. His name was Paul Gruchow, and one day he came to speak to one of my classes.

Gruchow owned a small bookstore in town, occasionally taught writing courses at our school, and had written several books of essays, one of which we'd read in our class. It was called *The Necessity of Empty Places* (St. Martin's Press, 1988), and I loved it for both the writing and the sentiment. I had no idea at the time that he had studied under poet John Berryman, or that for years he worked at newspapers and radio stations across the state before the University of Minnesota Press published his first book, *Journal of a Prairie Year*, in 1985 (it was reissued in 2009 by Milkweed Editions). All I knew was that his thoughtful, eloquent style had earned him comparisons to Thoreau and that somehow he had arrived at a place much like the one where I wanted to be.

After the class, I asked Gruchow if I could talk to him about writing. A few days later, he welcomed me into his home, told me to sit down, and offered me a cup of coffee. He was bald and portly and kind. His beard made him seem like the professor he sometimes was. He had a quick laugh and a look in his eye like his mind was always elsewhere.

We sat, and I started asking him how he'd done it, how it all went, what had been his first big break, and on and on. Patiently he told me about his work at the *Worthington Daily Globe*, about his first book, and about his many struggles along the way.

When I asked for advice, he tried to wave me off. He warned me that the writing life was full of hardship and disappointment and that there were seven times as many people who wanted to be writers as could be.

"Don't do it," he said, "unless there's nothing else you can do."

We sat for a long while, and I listened as he talked about his own writing life, hearing mostly the parts that I needed to hear. By the time I got up to leave, much of what he'd said had lodged itself deep into my mind. Before he'd even finished telling me his stories, I'd already begun to imagine my own.

* * *

I did not grow up in a storytelling family. My father tells what he likes to think are stories, but are more like sequential chains of loosely connected factual events. My mother keeps a three-line diary in which she catalogues the day's events, which is more like the raw material from which stories are made.

My wife's family, however, are easy raconteurs who tell stories loosely based on things that happened, but with deep feelings at their core. Her father, for example, likes to tell a story about how my wife's first car was a huge Lincoln Continental that was so big she could barely see over the dashboard—he could only see her little head in it. He got her that car, he says, because he wanted to make sure she was surrounded by as much steel as possible.

Except that wasn't it exactly. The Lincoln was just one of several cars her family owned and that she drove. Another was a tiny Datsun that would have been smashed like a tin can if it had hit another vehicle. Her first car was actually a crappy Ford Tempo.

For a long time I puzzled over this discrepancy. It took me years to finally understand that this wasn't really a story about her first car. It was a story about how much he loved his daughter and wanted her to be protected from the world. All that steel was love.

Why do we misremember things in certain ways? It's a fascinating question. Looking back, we do not recall a steady, seamless flow of events in time. Instead our mind breaks the flow of time into related chunks and stores them as scenes and anecdotes and episodes.

These episodes are the currency of our past and the storyboards we arrange to make sense of the things that have happened to us. We line them up like dominoes that lead to where we stand now. That we do this imperfectly has been written about many times. But I am more interested in the invisible threads running from one episode to the next, the forces that hold our stories together. Some have names, like love, or courage, or fear. Others are harder to pin down.

According to psychologist Dan McAdams, the episodes in our memory are not only the material for anecdotes to amuse our friends. They are also the building blocks of our "life story"—our own version of how we came to be the person we are.

Unless we write a memoir, or visit a therapist, we may never even tell anyone this life story, but that doesn't make it any less important. McAdams and others argue that the ability to see one's life as a story is at the heart of identity. In fact, our ability to "narrate" our life's events may even be the defining mark of consciousness.

Building a life story is a process that begins around the time we turn two years old. That's when we develop what McAdams calls a "primitive autobiographical self." As we move into adolescence, we start to emphasize different memories we feel were important—events in which we learned something or changed. Then, during our late teens we start to develop a more complicated "personal fable," in which we dream of the people we could become, like astronauts and presidents. McAdams calls this a "first draft" of our identity. We choose episodes based not only on who we think we are, but also on who we hope we can become.

As we move into young adulthood (between seventeen and twenty-five), things become a little more urgent as we try to compose a "full life story" that explains not only how we got wherever we are, but also what we believe, and who we will in fact be.

But our own past is not the only place from which our life story comes. The memories are our own, but what they mean and how we put them together come from the lives we see around us, from the stories we read and hear, and from whatever possibilities we can imagine.

For most of us, that full life story is never really finished, and is always subject to revision. Even so, it determines much of how our life unfolds. It's like a road map through the chaos, with arrows pointing one way or another at turning points like failure and success, death and birth, love and loss. That is what our daughter was really asking: *How do you live in a world with sadness and fear? And how should I?*

* * *

After the door closed behind me at Paul Gruchow's house, I went back to campus. I graduated and my career went slowly on. Yet even as I wrote story after story—hundreds of them—and even as I became a better writer, I still didn't quite know what a story was, not exactly. Instead I wrote by feel. A story was something I knew if I saw or felt it, but when I tried to put a definition into words, the meaning would slip through my fingers.

You can find this same problem running through much of the discussion about stories, or narrative, these days—and there's more of it than ever. Narrative neuroscience and narrative psychology are both growing fields. The Defense Advanced Research Projects Agency, a part of the U.S. Department of Defense, is even researching the use of narrative for defense purposes. Evolutionary biologist and author E. O. Wilson has repeatedly called us the storytelling species, and last April Houghton Mifflin Harcourt published a book by Jonathan Gottschall titled *The Storytelling Animal: How Stories Make Us Human*. But like other authors who have tackled the subject, Gottschall never quite articulates what he means when he talks about a story, and the book remains a disappointing collection of platitudes.

So what makes stories so important? What makes them stories at all? I finally stumbled across a kind of answer in a field about as far from the English department as you can get: artificial intelligence.

It turns out one of the biggest problems with making a computer intelligent is getting it to do something that we do naturally, something called "commonsense causal reasoning," which means understanding instantly when one thing causes another to happen.

"It's very simple things," says Andrew Gordon, a researcher at the University of Southern California's Institute for Creative Technologies. "Like if you tell the computer you dropped an egg, you want the computer to know that it broke, not bounced."

Gordon and his fellow researchers have been working on this problem for some time. They tried to instill this ability into a computer program by collecting millions of stories from blogs and using them to teach it how to deduce that A causes B.

After they had collected these stories, they designed a test in which they asked the computer a question, such as: "The man lost his balance on the ladder. What happened as a result? 1: He fell off the ladder. 2: He climbed up the ladder." Or this one: "The man fell unconscious. What was the cause of this? 1: The assailant struck the man in the head. 2: The assailant took the man's wallet."

"Computers are horrible at this test," says Gordon. Humans get the answer right 99 percent of the time—more or less perfectly. The best result they could get from the computer was 65 percent correct, or just 15 percent better than chance.

The computer, in other words, could not understand what we call *causality*. It couldn't see how the ripples spreading from one event caused another to occur. It couldn't see the forces that were secretly at work in our stories, but which we never name. For a computer, a Lincoln Continental is just a car—steel is just steel.

"Storytelling is a human universal," Gordon says. "There's not a culture that doesn't tell stories. It's something embedded in our genes that makes us good storytellers. It's a huge survival advantage, because you can encapsulate important information from one person to another and share it within a group. So there's a good reason to be good storytellers."

But the utility of storytelling has to do with causality, the ability to determine what causes what. Causality is the thing that helps you plan. Causality helps you decide what must be done to get what you need, or want, or want to avoid. You might know how the world is, but if you want to know how it got that way, you have to understand causality. If you want to know how to change it in order to effect your goals, or if you want to know what to expect in the future, you have to understand causality. When you tell a story, you're trying to bring what Gordon calls "causal coherence" to events that are ordered in time.

Whether computers will ever be able to understand not only what happens in a story, but also why it happened and why it matters, remains uncertain. At the moment, they are very far from that point.

We, on the other hand, are already there. We see causality constantly, incessantly, and effortlessly: when we read the news, when we gossip about neighbors, when we watch a movie or read a book.

Much of our life is the search for the causal links between events, for the forces at work not only in the physical world, but also in the hearts and minds of the people we know. We are constantly cataloguing the story lines around us in an effort to sort out our own. What causes greatness? What causes failure? What causes happiness? What causes goodness or evil? What causes sadness and fear?

Radio journalist Ira Glass has said that his mentor, Keith Talbot of National Public Radio, once advised, "Every story is an answer to the question: How should I live my life?"

Or, as the poet Muriel Rukeyser once wrote, "The universe is made of stories, not atoms."

* * *

The headline came as a shock: "Author Paul Gruchow, who chronicled the prairie, dies at 56." In late February of 2004, Gruchow took his own life with a drug overdose.

There were few details. Obviously, he had been deeply depressed. According to one article, when asked several months before his death how he wanted to be remembered, Gruchow replied, "Tell them I got up and said a few words." According to another, when an old friend wrote to ask if he could do a story about him, Gruchow wrote back: "Last year I earned $62.85 in royalties and gave one public talk, in Duluth, that drew a dozen listeners. . . . Two or three times the phone rings. Usually I don't answer it. There isn't a story."

There was a story, but perhaps not one he wanted to tell. It almost certainly wasn't the one he'd imagined when he dreamed of becoming a writer. Maybe it was the story he'd been trying to tell me all those years ago when I sat across from him.

But it wasn't the story I heard. What I heard was that it was not going to be easy, that it would take time and effort, and that I would have to endure hardships. Those were warnings that have served me well.

Looking back now, his words seem to take on another meaning, another kind of caution, one that has little to do with writing, and everything to do with life: Down there in our stories, the ones we tell ourselves, the ones we tell others, the ones we hope are true, the ones we fear might be, are forces at work that we can only ever halfway understand. Knowing how causalities hold our past together doesn't mean we can always see what those causalities are.

What I heard from Gruchow was this: Writing, creating something so beautiful that it may outlast you, is so important that you must be prepared to suffer for it, and then keep going on. That has always been a part of my story, and that is one of the reasons I am still writing nearly twenty years later.

That may also be why the news of Gruchow's death, so many years after we met, filled me with a deep and unexpected sadness. It was a sadness born of the realization that while I thought he and I had been reading from the same script, perhaps we weren't. It drove home the understanding that at each of life's crossroads, what you believe deep down determines which way you turn.

Be mindful, in other words, of the stories you believe, the stories you love, and the stories you choose to tell. Because in the end they may become your own.

16

Conversation in Intensive Care

Amy Wright

Mind ceased to be an abstraction for me, when at twenty-eight a car accident broke the glass of the driver side window against my skull. After I regained consciousness for enough days that the neurology team could risk an angiogram, I peered inside my fractured cranium. Blood vessels supplying oxygen to my brain throbbed, lit by radioactive dye, a bioluminescent jellyfish pinned inside a cave after the swollen tide had ebbed.

Though the image on the screen appeared isolated, it streamed from a camera inserted in my groin and unreeled like fishing line. If yoga practice had not already taught me the yoke between mind and body, the tether of my femoral artery would have assured me that the brain is not alone in composing mind.

Scientists in the past decade have conducted numerous studies into the mind-body connection, often citing the "gut-brain axis" that regulates hunger and other aspects of homeostasis, mental health, and intuition. I grew interested in this research after I recovered and wanted to know more about the injured regions of my brain. I started reading journals like *Nature, Psychological Bulletin,* and the *Journal of Comparative Neurology*. These articles helped explain why my memories from those weeks in the hospital and afterward were primarily implicit and visual. Areas associated with auditory and verbal short-term memory and language processing had been slammed into the cranial cavity during the impact. Other areas, associated with motor control, behavior, memory, judgment, and "giving meaning to bodily states," were further from the trauma and responded better after the hemorrhaging subsided.

One of the articles I read, by Canadian neuropsychologist Simon McCrea, described a form of cognition we don't usually associate with thinking, known as "nonverbal encoding."[1] We use this ability every time we recognize facial expressions or spot a red flag in someone's body language, such as avoidance of any eye contact, but the term gave me a way to consider the intuitive forms of cognition that remained active in me, which I credit with certain aspects of my recovery.

Verbal recoding

"Someday soon," John D'Agata says, "we're going to have to divorce our understanding of *essaying* from our understanding of the thing we'll start to call nonfiction." He attributes this need to the end of the Second World War, the birth of New Journalism, and the widening gap in the mid-twentieth century between reality and humans' ability to comprehend the reality of Nazi death camps in Europe, Japanese-American internments in America, the Nanking Massacre in China, and other atrocities, even as jets, the microwave oven, and the computer are invented. But this divorce was not a one-time generic split occurring *ca.* 1945. It is the often-prolonged process by which essaying solidifies into realization, and it is as entangled as the marriage between brain and body.

D'Agata does not distinguish nonfiction from creative nonfiction in *The Lost Origins of the Essay*, but in divorcing the activity of essaying from the forms nonfiction takes he calls attention to how this genre illuminates the mind at work. It may be nonfiction's primary function, as Montaigne suggests in *Essais*, "to follow a movement so wandering as that of our mind,"[2] except *our* mind rarely comes into accord between even two points of view. Discord seems fundamental to this process, since argument drives the dialectic Montaigne follows between his aging body and the mind that once mastered it, which it now "dominates in turn." But, argument is limited. To wander freely, the mind cannot be reduced to reaction equal only to its opposition; it must be available to another form of communication altogether.

Consensus

"Are you warm enough?" a nurse asked when I shivered while being transported from my hospital room to radiology. I rolled my head for a few degrees on the pillow, and she summoned an extra blanket. Then, a ring of arms lifted me from the gurney to the examination table. As during a high fever, attention zeroed in on sensation. Palpating scopes sprouted from each nerve ending. Mind became a field of sense whose grasses swayed past the body's contours. The primary conversation was taking place between members of the medical team, but I was heeding the exchange between a sheet of Fentanyl and a throb of pain.

Neurologists had previously conducted several CT scans of my brain while I was unconscious, but I was conscious for the angiogram. I turned my head to watch the screen glow with an organ that would define my future, although I could not conceive of a future yet. Thankfully, I think now, I did not fear the worst or hope for the best. I responded to the medical team, my parents, and visitors, but I was not in the usual dialogue with myself. The internal back and forth that assesses, argues, and corrects had gone into radio silence. I was no more an arbiter of differing points of view, cultures, eras, and ethical models. I narrated no storylines. I was an unqualified state of being, covered by a white, waffled blanket steeped in the chrome and rubbing alcohol smell of the room.

Decoupling

D'Agata points to Paul Celan's "Conversation in the Mountains," as an example of the pending separation of essaying and nonfiction, because Celan calls on dialogue, that age-old essay form, to reckon with different voices within himself. What results is a part-narrative, part-lyric address to that gulf.

Celan's essay opens one evening when a Jew "with an unpronounceable name" leaves his house, making himself heard with his walking stick knocking against the stones. "You could hear it," his narrator invites readers to imagine in the first paragraph-long sentence. The sun is setting, and the stick seems to tap out as if in Morse code, "do you hear me, you do, it's me, me, me and whom you hear, whom you think you hear and the other." In decoding this hidden declaration, Celan identifies an inner voice, a fabricated voice, and an "other" who does not speak in the same way. The trialogue becomes clearer later in the essay, when the Jew meets his cousin on the road.

His cousin asks why he's traveled so far, and he answers that he needed to talk with more than his stick, which talks only to the stones. His cousin chides him that stones do not talk, they speak, and what they say they say to no one. They hold a conversation between "nobody and Nobody," he says, beyond mouth and tongue. The cousin's awareness of the stones' speechless-speech makes his own speech ironic. For all his understanding, he can do little more than decry its limitations.

Celan did not anticipate his words resonating with others across time, only that words can easily be mistaken. The narrator laments all this talk, pitying the two men their missed opportunity to experience something more profound. A kind of superego, he criticizes their preoccupation with words and notes their failure even to notice the maiden-pink blooming at their feet, which they later acknowledge not seeing in the fullest sense. "I see it and don't see it," one admits to the other: "In my eyes there is that moveable veil, there are veils . . . you lift one, and there hangs another." The three of them—the Jew, the cousin, and the narrator—layer one another and the world around them with projections and judgments. They recognize that they are doing it, in turns, but their recognition resolves nothing. They continue to blame one another, culture, human nature, and personal weakness for not meeting life without any blinds. The "windbags" rib each other good-naturedly but acknowledge that this predicament keeps them from living and loving freely.

Conversation in the mangroves

Who loves you? Tessa Fontaine wants to ask everyone in "The Gators Have Turned." In this personal essay, Fontaine has been losing her mother to a massive stroke for the past two years, so her question is poignant with wanting to reach out to others. But the question isn't in the first draft, and presumably not consciously in the experience itself, where it may have bolstered some support. It is a later insight into loss that was occurring on another level—before language and in the body—which she accessed through the acts of essaying and grieving. Looking at the forms this process takes over several drafts reveals the particular lens nonfiction shines into the mind's complexity.

Fontaine shared several pre-published versions of this narrative with me, when I expressed interest in how essays turn to circulate the known with the unknown. In all of these versions, she is riding a tin motor boat through the Delta, listening to the boat's driver, Leila, talk about how the alligators in Mississippi have turned on one another. In the earliest draft, Leila also tells Fontaine the story of a five- or six-year-old local girl named Caroline, who didn't come one day when she was called in for lunch. Family, neighbors, and authorities scoured the area desperately for eleven days after she went missing. Just when they'd resigned themselves, distraught, to the likely fate of little girls alone in the swamps, Caroline showed up "in new, clean clothes, well-fed, rosy-cheeked and happy." After their elation subsided, her parents asked where she had been, and she told them about a little girl who had taken care of her. The police asked her to take them to see the little girl, and there they found the bones of a child who disappeared eighteen years before. "Oh, yes," Caroline said, "Yes, the little girl was dead."

The uncanny anecdote doesn't make it into the essay published in *Seneca Review* or the sections from it that appear in *The Electric Woman: A Memoir in Death-Defying Acts* (Farrar, Straus, and Giroux). Perhaps in the context of her mother's stroke it seemed sunny to suggest someone taking care beyond the grave, but the fact that Fontaine first heard Caroline's story in that context remains fundamental to this act of essaying, and the nonfiction on which it left its trace.

The community that gathers to search for any missing person is one answer to the question "Who loves you?" Fontaine may in fact have been inspired to ask it by imagining Caroline's surge of support. What is certain is that the question rather than the story remains, and both emerge from her determination to follow the boat's driver "as deep into this wild as she'll take me." This motive, present in every version I read, echoes Montaigne's desire to follow the wandering mind and offers an illustration of that route, since mangrove roots present physical obstacles and house mysteries at every turn.

Following the mind means tracing the known into the unknown, but the mystery of Caroline's return makes clear that essaying is in large part a reckoning with the unknowable. Even together there are things we cannot know, but sometimes, alone, by innocence or accident, we glean an understanding that lies beyond.

Roundabout

The angiogram revealed where a hematoma, or internal blood clot, had formed in my mid-cerebral artery and stretched its walls like a blown-up balloon. It also confirmed my neurologist's hope that the blood had reversed directions and took an anatomical detour known as the Circle of Willis.

Named after London physician Thomas Willis, the Circle of Willis is a ring of arteries at the base of the brain around the pituitary gland. When the flow of blood in this area is obstructed for any reason, the arterial loop provides an alternative. Others before Willis had noticed this feature, but he was the first to realize its function in his illustrated *Cerebri Anatome*, which in 1664 laid the foundations for neurology. What he did not realize is that only about a third of the population has this option. A complete circle exists in most people, but the infrastructure connecting these arteries to the rest of the brain is often incomplete.

Had my brain never called on my Circle of Willis, its maintenance of that supplementary network might seem akin to a cautious mom who carries an EpiPen for a child with no known allergies. That it did was my incredible fortune. But that it exists at all bespeaks the wonder of evolution, which protected this path over generations of ancestors who suffered falls or blows to the head and lived. Such an intelligence demonstrates fascinating agency, but can one follow that aspect of mind?

It leads to a realm as wordless as the artery where my blood rushed and stopped, for there are events we lack the sensitivity to sense. Yet, humans have invented instruments to enter some of those moments. They magnify texts written on the leaves of bones, frame abstracts of gastric juices, swipe clues under glass. They probe cells that speak only to their walls but convey a system that reaches and splits, venturing ever forward until it needs to double back.

Between thought and sense

When her uncle Leland died, Rebecca McClanahan says she was surprised at the depth of her grief. Mourning the last of her uncles, she went in search of poems and essays about nieces and uncles to comfort her and help her understand this loss. She found works by nephews about their uncles and by nieces about being abused by their uncles, but she realized that she would have to write about the particular protection her uncles offered her.

The effort had begun years before, after previous losses, with a poem divided into sections devoted to each uncle, but she realized that she needed more room and prose to address this grief. In the resulting essay, "The Uncles," she lets the stories of her uncles speak to one another. While Ivan had a job raking rivers and Merrill raised prize rabbits as a boy, Leland was a farmer with a left eye that wandered away from conversations. One year when the soybean crop brought in extra money, Leland bought a Hammond organ and taught himself to play turn-of-the-century love songs. McClanahan's affection for all of her uncles is clear, but she admits that she knew them less intimately than her aunts, whose beds and toilet she sometimes shared and whose recipes and china she inherited.

But, she says, the idea for this essay did not emerge until she had a dream—that form of communication by which the mind lets slip one of its veils. It revealed to her that while the deaths of her aunts opened a space around her, as if they had saved room for her to grow into, her uncles' deaths make her feel diminished, like a child again.

In the dream described in the essay, all of her uncles return and one by one place their heavy hats on her head. She pulls their dark overcoats over her arms and slips inside. Their hats slide down over her eyes, and she catches "the reservoir of tobacco and motor oil and freshly cut meadow grasses." These remembered smells function like a cellular tower connecting one part of her mind to another, conveying a sense that supersedes thought, for which the body is a bioelectromagnetic antenna.

Circumlocutionary system

Fifteen years after the accident, I sat at my desk watching NASA recordings of a solar storm that stretched from mid-October to early November. Named the "Halloween Storms of 2003," this solar activity generated the largest coronal mass ejection since solar storms have been recorded, but its measure can only be estimated at somewhere between seven and twenty-five times more powerful than the largest recorded flare, because its force overloaded NASA and Geostationary Operational Environmental Satellites (GOES) equipment.

The night a hatchback slammed into my jeep, and my jeep's roof slammed into a tree, one of the storm's smaller flares, still thirteen times the size of Earth, leapt from the Sun into the darkness of space. I replayed the video of it over and over. A flame of radiation tore away from the corona and dove headlong into nothingness, like the photograph of the artist Yves Klein donning a business suit, outstretching his arms, and plunging chin-up, lips sealed into the void.

Watching, I wondered if I flinched or cried out when the hatchback's tires shrieked toward my passenger side. I know muscle memory records some of our most basic tasks, but did my stereocilia, those tiny hairs in the inner ear, retain the racket of the hit? I was not the primary actor on scene, slumping into the cradle of my seat belt when the jeep thundered to the ground. Maybe it seems absurd to query an unconscious form. It makes more sense to transfer agency to the collective that rushed around me, the emergency medical personnel who staunched my wounds and inserted oxygen tubes. And, I might better acknowledge the grace by which I survived, but faith is defined by unknowability. I can better attempt to access the experiential, nonmystical divinity of the body.

Had I been able to gaze up from that corner between Clarkson and Evans Street, I could not have seen the solar storms. The moon did not mirror them. Even were the sun directly overhead, my eyes would have been poor scopes for 92 million miles. But the silence of the video prompted me to listen harder to what remained of my own footage, and I picked up a signal from the only internal line of communication that continued to hum.

Like Celan's stones, this internal line did not talk but spoke. It declared with its whole being that it is not what debates existence.

Even its calm accord insisted nothing. Although it could not lift a finger to pluck glass from its eyelid, it waited. Breathing shallowly, it inventoried. It turned on blood vessels like sprinklers and soaked the left hemisphere. Without deliberation or pity, it daubed the gush with tissues and lowered the blood pressure to a walking bass line. Going about its work neither admired nor criticized, it tied a knot of blood cells into a parcel. It herded oxygenated cells like drenched children out the back door. A caretaker taking care, it launched endorphins like sky lanterns, fluffed the cranial cavity with cytokines, noticed the off-kilter angle of the ankle, and shaded the eyes.

Then, the whole sea sloshed and sirens wailed. It heard them crying out but was not tempted, too busy numbering every splinter of bone and hair on this head. The ears were ringing, cymbals crashing like ships into rocks, but far below them an anchor the heart had dropped in the pelting rain held. A flock of hands descended on the floodlit surface and began to carry away the metal, the glass, and the woolen overcoat. The night was cold, it says, but there was a stone in the pocket of its stomach, warm and pulsing with a peace that surpassed all understanding.

17

Mindfulness and Memoir

Julie Wittes Schlack

Our capacity to see

"The extended self, which is what we think of as our self, is essentially a story," declares neuropsychologist Paul Broks.[1] "It's the story of what has happened to this body over time."

Broks could also be defining autobiography. But memoir is something a little different. Memoir's purpose isn't to recount so much as it is to reflect. Great memoir depicts experiences and events, yes, but according to author and critic Sven Birkerts, in writing it, the author's purpose is "to discover the nonsequential connections that allow those experiences to make larger sense; they are about circumstance becoming meaningful when seen from a certain remove.[2]" And in writing memoir, the author is keenly aware of what Jon Kabat-Zinn, Professor of Medicine and founder of the Mindfulness-Based Stress Reduction Clinic, describes as "seeing as a relationship, the relationship between our capacity to see and what is available to be seen."[3]

As I've concurrently studied the practice and philosophy of mindfulness meditation and the discipline and great exemplars of memoir, I've been struck by the similarities in the two processes and in their inherent philosophical and psychological challenges.

In her landmark essay "Memory and Imagination," writer Patricia Hampl describes the central challenge of the mind in relation to the self, addressing from a literary perspective what Jon Kabat-Zinn, Dan Siegel, Paul Broks, V. S. Ramachandran, and others explore in the realms of philosophy and neuroscience:

> "Each of us must possess a created version of the past. Created, that is, real in the sense of tangible, made of the stuff of a life lived in place and in history. And the downside of any created thing as well: We must live with a version that attaches us to our limitations, to the inevitable subjectivity of our point of view."[4]

Though Hampl (and memoirists in general) address themselves to the past, the same challenge holds true in both experiencing and reflecting on what's occurring in the present. So it strikes me that experts from each of these disciplines are asking the same essential question: *How do we make sense of our own experience without becoming a prisoner of our own narrative, our own construction?*

I won't presume to offer a definitive answer, but I'll explore the question itself in depth, from multiple perspectives, by focusing on four concepts from the realms of mindfulness and neuroscience that have direct application to memoir writing and reading:

- Dual awareness and the quest for non-dual awareness, or what Jon Kabat-Zinn describes as "[the] lack of boundary between knowing and the known, seeing without a seer";[5]

- Our sense of self as something real but constructed, as continuous and self-contained but endlessly dynamic;
- Our sense of time passing, which accelerates and decelerates as a consequence of age, cognitive development, and our (in) ability to be present in the moment;
- Our need to create a coherent life narrative and the companion danger of investing in a story that obstructs our awareness of what is transpiring in the moment.

Dual awareness

"Memoir is the intersection of narration and reflection, of storytelling and essay writing," writes Patricia Hampl. "It can present its story *and* consider the meaning of the story."[6] Consider this definition alongside Buddhist and mindfulness scholar Jon Kabat-Zinn's observation, "We grow and change and learn and become aware through the direct apprehension of things through our five senses, coupled with our powers of mind, which Buddhists see as a sense in its own right."[7] Both the writer and the Buddhist are describing a duality—the act of sensing and the act of observing oneself doing so and interpreting the experience.

In memoir, those two processes—sensing and reflection—are often separated in time, where the experience is remembered rather than felt in the moment. The author straightforwardly tells the reader what she recalls from the past and what she realizes now. The act of considering and interpreting the sensory experience is deferred and rational, as in this example from Joan Didion's essay, "On the Morning after the Sixties."[8]

> I was lying on a leather couch in a fraternity house (there had been a lunch for the alumni, my date had gone on to the game, I do not

> now recall why I had stayed behind), lying there alone reading a book by Lionel Trilling and listening to a middle-aged man pick out on a piano in need of tuning the melodic line to "Blue Room." . . . That such an afternoon would now seem implausible in every detail—the idea of having had a "date" for a football lunch now seems to me so exotic as to be almost czarist—suggests the extent to which the narrative on which many of us grew up no longer applies.

Memoirists also juxtapose the then and the now, the sensation and the later understanding of it, simply by using a word or displaying an intellectual sophistication that wasn't available to the author then, but is only now. Look at this example from Vivian Gornick's *Fierce Attachments*, as she tells us about the apartment in which she grew up:

> The clear air, the unshadowed light, the women calling to each other, the sounds of their voices mixed with the smell of clothes drying in the sun, all that texture and color swaying in open space. I leaned out the kitchen window with a sense of expectancy I can still taste in my mouth, and that taste is colored a tender and brilliant green.
>
> For me, the excitement in the apartment was located in the kitchen and the life outside its window. It was a true excitement: it grew out of contradiction.[9]

The use of the word "still" ("a sense of expectancy I can still taste in my mouth") is all that's needed to tell us that significant time has elapsed between the vivid sensory experience and the act of writing about it. Even more artful is the last sentence ("It was a true excitement: it grew out of contradiction."), where Gornick presents us with an unquestionably adult understanding of a childhood emotion.

The dual awareness to which Kabat-Zinn refers is most thrillingly displayed when the past and the present, the child and the adult, the experienced and the understood are even more proximate, when they are simultaneous, as in this excerpt from *Fierce Attachments*:

> She holds tightly to my arm. She neither confirms nor denies my words, only looks directly into my face. "Remember," she says. "You are my daughter. Strong. You must be strong."
>
> "Oh Ma!" I cry, and my frightened greedy freedom-loving life wells up in me and spills down my soft-skinned face, the one she has given me.[10]

In this sequence of excerpts, we see the gap between observed and observer narrowing, and the prose becomes more powerful as a result. Even when describing the past, to achieve such detailed, vivid memories Gornick must essentially re-live that experience, which begs the question of whether imagination is, in essence, the ability to be present in a different time or place, to occupy the story one is constructing.

But Kabat-Zinn challenges us to dissolve that barrier entirely:

> We never dream that there may be observation without an observer, that is, until we naturally, without any forcing, fall into observing, attending, apprehending, knowing. In other words, until we fall into awareness. When we do, even for the briefest of moments, there can be an experience of all separation between subject and object evaporating. There is knowing without a knower, seeing without a seer, thinking without a thinker, more like impersonal phenomena merely unfolding in awareness.[11]

What is so startling in Kabat-Zinn's description is how closely it parallels memoirist Jill Bolte Taylor's account of what it was like to have suffered a massive stroke, one in which the functioning of her left cerebral hemisphere—site of one's ability to process language—was shut down, and her consciousness was informed solely by her right hemisphere.

> I essentially became an infant in a woman's body . . . in that moment, my brain chatter—my left hemisphere brain chatter—went totally silent. Just like someone took a remote control and pushed the mute button. Total silence. And at first I was shocked to find myself inside of a silent mind. But then I was immediately captivated by the magnificence of the energy around me. . . . Because I could not identify the position of my body in space, I felt enormous and expansive, like a genie just liberated from her bottle. And my spirit soared free, like a great whale gliding through the sea of silent euphoria. Nirvana. I found Nirvana.[12]

In essence what Bolte Taylor is describing is a permanent present, and one in which there are no boundaries. The sense of "self" as an object blurs into a broader field of energy and sensation. But while being perpetually in the now may benefit us as humans on a path to Nirvana, it eventually wearies readers. With only a continuous now, there is no narrative arc. And without the boundary of "self" creating a tension between an author's interior and exterior lives, there is little to engage us.

Our sense of self

Neuroscientist Paul Broks writes that:

> The illusion is irresistible. Behind every face there is a self. We see the signal of consciousness in a gleaming eye and imagine some ethereal space beneath the vault of the skull, lit by shifting patterns

> of feeling and thought, charged with intention. An *essence*. But what do we find in that space behind the face, when we look?
>
> The brute fact is there is nothing but material substance: flesh and blood and bone and brain. I know, I've seen. You look down into an open head, watching the brain pulsate, watching the surgeon tug and probe, and you understand with absolute conviction that there is nothing more to it. There's no one there. It's a kind of liberation.[13]

I use the term "self" as though it is something as concrete, tangible, and commonly understood as "rock" or "slipper." But Kabat-Zinn argues, "There is no permanent, isolated, self-existing self. . . . Both the sense of personhood and our personality are in a profound way impersonal, although clearly unique and relatively real. . . . We need to question whether the sense of self is fundamentally real or just a construct of the mind."[14]

On this point, many memoirists would likely agree: the "self" *is* a construct of the mind. But here is where some principles of "good" mindfulness versus good memoir diverge, because that "sense of self" is as essential to vivid and compelling storytelling as the other five perceptually based senses. Neurologist V. S. Ramachandran says that the sense of self rests on seven pillars,[15] and I'd argue that good memoir does too.

(1) **Unity**: Despite the diverse and continuous onslaught of sensation that you experience from one moment to the next, Ramachandran notes that "you feel like one person. Moreover, all of your various (and sometimes contradictory) goals, memories, emotions, actions, beliefs, and present awareness seem to cohere to form a single individual."

Memoir is made of goals, memories, emotions, actions, and beliefs; they are the bricks that the writer carefully selects and aligns. Of course what is so fascinating is that over time, the memoirist may erect different buildings from the same bricks—a fortress in one book, a maze in another—or choose an entirely new set of construction materials, as author Mary Karr observes when she writes:

> Though *The Liar's Club* rang true to me when I wrote it, from this juncture it seems to have sprung from a state of loving delusion about my family. In those days, I still enjoyed a child's desperate tendency to put sparkles on my whole tribe. Were I writing that story today, I'd be less generous to them while perhaps shining more empathy on my younger self. . . . The self who penned that book formed the filter for those events. I didn't fabricate stuff, but today, other scenes I'd add might tell a less forgiving story.[16]

These differences reflect the fact that memoir is fundamentally an act of interpretation. The present "me" examines the past "me," acutely aware of the differences between the two borne of age and life events, but nonetheless experienced as a single, unified self.

(2) **Continuity:** Despite the distinct events punctuating your life, you feel a continuous identity through time, and can easily navigate from early childhood memories to projections of your own future.

If "unity" comprises the bricks from which memoir is assembled, "continuity," or what Sven Birkerts describes as "the need to plait together past and present," is the mortar. "I now, I then" is the very essence of continuity and of memoir, as Birkerts illustrates with this excerpt from Virginia Woolf's "A Sketch of the Past"[17]:

> 2nd May . . . I write the date, because I think I have discovered a possible form for these notes. That is, to make them include the present—at least enough of the present to serve as platform

> to stand upon. It would be interesting to make the two people, I now, I then, come out in contrast. And further, this past is much affected by the present moment. What I write today I should not write in a year's time.

If Woolf explains the technique, memoirist Abigail Thomas executes it stunningly in *A Three Dog Life*:

> This is the one thing that stays the same: my husband got hurt. Everything else changes. A grandson needs me and then he doesn't. My children are close then one drifts away. I smoke and don't smoke; I knit ponchos, then hats, shawls, hats again, stop knitting, start up again. The clock ticks, the seasons shift, the night sky rearranges itself, but my husband remains constant, his injuries are permanent. He grounds me. Rich is where I shine. I can count on myself with him.[18]

(3) **Embodiment:** This pillar refers to feeling anchored and at home in your body in the most literal way—"you know that your hand belongs to you and that the waiter's hand doesn't."

Not only does physical sensation play a prominent role in memoir, but authors often describe the sensation of feeling at home in their bodies with language like "coming into my own," and usually only arrive there after painful adolescent years of frequent betrayal by their constantly changing and unpredictable bodies. Robust embodiment—not just knowing the distinction between one's own hand and someone else's, but feeling in control of one's body—seems to be a prerequisite to the more developed and metaphysical sense of self.

(4) **Privacy:** Ramachandran asserts, "Your qualia and mental life are your own, unobservable by others." You can empathize with others, but not really experience what they experience.

Recent research into the action of mirror neurons—brain cells that respond the same way to witnessing someone else perform an action as they do when you perform that action yourself—suggests that Ramachandran's construct may require some tweaking. It appears that humans may minutely ape the facial expressions of others in order to evoke what they are likely feeling. By copying another person's physical response to a stimulus (whether internal or external), we do a pretty good job of recognizing what caused that response. In short, mimicry is the first step to building empathy.

But empathy does not preclude privacy. Even when we understand and feel the feelings of others, we also observe ourselves doing so, judging the validity of those feelings, anticipating their social desirability, and engaging in all sorts of internal chatter that is not shared.

For example, listen to Anne Lamott write about the true, shameful feeling of jealousy when a fellow writer meets with great commercial success, especially if that person is a friend.

> You are going to feel awful beyond words. . . . It can wreak just the tiniest bit of havoc with your self-esteem to find that you are hoping for small bad things to happen to this friend—for, say, her head to blow up. Or for him to wake up one morning with a pain in his prostate, because I don't care how rich and successful someone is, if you wake up having to call your doctor and ask for a finger massage, it's going to be a long day. You get all caught up in such fantasies because you feel, once again, like the kid outside the candy-store window, and you believe that this friend, this friend whom you now hate, has all the candy. You believe that success is bringing this friend inordinate joy and serenity and security and that her days are easier.[19]

As socially undesirable as it is, jealousy is a creaky, echoing haunted house that's actually built on a foundation of empathy with someone else's happiness. And that's precisely why we tend to keep it locked away as much as possible, visible only to our own, scolding selves.

(5) **Social embedding:** Ramachandran writes that "almost all our emotions make sense only in relation to other people . . . the self needs to feel part of a social environment that it can interact with and understand on its own terms."

In essence, he is saying that the sense of self depends on its own juxtaposition with a broad "other" to exist. The central thread of most memoir—particularly the coming of age story—is precisely that quest to move from "me vs. the world" to "my place in the world." In memoirs such as Tobias Wolff's *This Boy's Life*, Mary Karr's *The Liar's Club*, or Henry Roth's *Call it Sleep*, the world that these authors interact with and try to understand largely comprises their immediate family. In the work of essayists like Joan Didion, the challenge the author works through on the page is to understand her own experience within the context of societal events and upheavals, as in this excerpt from *The White Album*:

> Driving a Budget Rent-A-Car between Sacramento and San Francisco one rainy morning in November of 1968 I kept the radio on very loud . . . in an effort to erase six words from my mind, six words which had no significance for me but which seemed that year to signal the onset of anxiety or fright. The words, a line from Ezra Pound's "In a Station of the Metro," were these: *Petals on a wet black bough*. The radio played "Wichita Lineman" and "I Heard It through the Grapevine." *Petals on a wet black bough.* Somewhere between the Yolo Causeway and Vallejo it occurred to me that during the course of any given week I met too many people who spoke favorably about bombing power stations. Somewhere between the Yolo Causeway and Vallejo it also occurred to me that the fright on

> this particular morning was going to present itself as an inability to drive this Budget Rent-A-Car across the Carquinas Bridge. *The Wichita Lineman was still on the job.* I closed my eyes and drove across the Carquinas Bridge, because I had appointments, because I was working, because I had promised to watch the revolution being made at San Francisco State College and because there was no place in Vallejo to turn in a Budget Rent-A-Car and because nothing on my mind was in the script as I remembered it.

Without that tension between internal and external reality, the memoirist would have little to say.

(6) **Free will:** You can envisage different courses of action and have the sense of making deliberate choices.

Though the belief in free will and self-determination has taken on religious overtones in American culture, becoming something of a self-serving myth, it is nonetheless true that the ability to envisage different choices and outcomes, and then to choose the more desirable path and act on it, is a defining feature of the human brain. It is also a defining theme of memoir. From choosing who to play with or who to marry, which secrets to reveal and which to harbor, when to leave home and when to return to it, the heart of much memoir is in chronicling the choices one has and has not made, and in reflecting on their consequences.

(7) **Self-awareness:** Ramachandran writes, "A self that is not aware of itself is an oxymoron . . . your self-awareness might partly depend on your brain using mirror neurons recursively, allowing you to see yourself from another person's (allocentric) viewpoint. Hence the use of terms like 'self-conscious' (embarrassed), when what you really mean is being conscious of someone else being conscious of you."

Self-awareness is empathy in reverse; it's the capacity to see yourself as you imagine others see you. That ability gives rise to the internal conflict that lies at the racing heart of many memoirs, particularly those focused on the author's adolescence and early adulthood. That gap between how we experience ourselves and how we think others experience us forms a trapdoor into a state of longing, aspiration, and rich confusion. And when employed by a mature memoirist, one who is cognizant of their role in life as a digger, harvester, and observer, this almost meta-self-awareness is especially rich. Witness this passage by David Sedaris in *Repeat After Me*, writing about his relationship with his sister (and her parrot Henry):

> She's afraid to tell me anything important, knowing I'll only turn around and write about it. In my mind, I'm like a friendly junkman, building things from the little pieces of scrap I find here and there, but my family's started to see things differently. Their personal lives are the so-called pieces of scrap I so casually pick up, and they're sick of it. Our conversations now start with the words, "You have to swear you will never repeat this." I always promise, but it's generally understood that my word is no better than Henry's.[20]

Without self-awareness, we could not vividly envision and depict others or ourselves. We could not create art.

Our sense of time passing

In *Coming to Our Senses*, Kabat-Zinn explores the "sense" of time passing. He alludes to computer scientist Ray Kurzweil's theory that "our internal, subjective sense of time passing is calibrated by the interval between what we feel or sense as 'milestone' or noteworthy events, along with 'the degree of chaos' in the system," and goes on to note:

> Babies and young children have lots of milestone events happening in those formative years and the frequency of such events decreases over time, even as the level of chaos in the system (say, for example, unpredictable life events) increases. The interval between milestone events is short and thus the felt experience of childhood is one of timelessness, or of time passing very slowly. We are hardly aware of it, we are so much in the present moment. As we get older, the spaced intervals (time) between noteworthy developmental milestones seems to stretch out more and more. . . . Subjectively, it feels like time is speeding up as we age because our reference frame is growing longer."[21]

Anyone over the age of forty probably recognizes this phenomenon of time seeming to accelerate, and of feeling that our lives have comprised two distinct phases: *childhood*—long, languorous, and rich with sensation and detail; and *everything that's happened since*, an exhausting blur of teachers, bosses, bills, cars, boxes packed and unpacked, debts acquired, some of them paid—all punctuated by a few key events like weddings, births, and funerals.

But the fluid and fairly constant shuttling between the past, the present, and the imagined (or implied) future is what transforms chronicle into memoir. Knowledge of the past and projections about the future lend essential richness to the literary (and, I would argue, lived) experience, as described by Virginia Woolf in "A Sketch of the Past":

> I was thinking about Stella as we crossed the Channel a month ago. I have not given her a thought since. The past only comes back when the present runs so smoothly that it is like the sliding surface of a deep river. Then one sees through the surface to the depths. In

> those moments I find one of my greatest satisfactions, not that I am thinking of the past; but it is then that I am living most fully in the present. For the present when backed by the past is a thousand times deeper than the present when it presses so close that you can feel nothing else, when the film on the camera reaches only the eye.[22]

Slowing down the felt sense of time passing, living most fully in a present that is "backed by the past," seeing "through the surface to the depths"—these are the most challenging tasks for the writer, and it is here that mindfulness practice perhaps has the most to offer. "The other way to slow down the felt sense of time passing is to make more of your ordinary moments notable and noteworthy by taking note of them," Kabat-Zinn writes. "The tiniest moments can become veritable milestones. . . . Since there are an astronomically large number of moments in the rest of your life, no matter how old you are, the more you are here for them, the more vivid life becomes."[23] Taking note of ordinary moments—a skill so essential to memoir—is a discipline that mindfulness meditation reinforces kinetically through the practice of attending to one's breath when meditating.

But note that Kabat-Zinn doesn't say the "happier" or "more serene" life becomes as a consequence of meditating. Indeed, "vivid" is precisely the right word for what both meditators and writers aspire to. In great memoir, the past and the present collide and converge, enabling moments to be more deeply felt and specifically rendered, as in this poignant excerpt from *A Three Dog Life*, Abigail Thomas's memoir about her husband's severe brain injury:

> Rich is lodged in a single moment and it never tips into the next. Last week I lay on his bed in the nursing home and watched him. I was out of his field of vision and I think he forgot I was there. He stood still, then he picked up a newspaper from a neat pile of newspapers, held it a moment, and carefully put it back. His arms

> dropped to his sides. He looked as if he was waiting for the next thing but there is no next thing.
>
> I got stuck with the past and future. That's my half of this bad hand. I know what happened and I never get used to it. Just when I think I've metabolized everything I am drawn up short. "Rich lost part of his vision" is what I say, but recently Sally told the nurse, "He is blind in his right eye," and I was catapulted out of the safety of the past tense into the now.

Awakening to the moment when "before" converges with "now"—what Kabat-Zinn describes as "falling into awareness"—isn't always pleasant. But it is essential to what Buddhist writer and teacher Natalie Goldberg advises memoir writers to do: "Build a tolerance for what you cannot bear. This is good practice. It makes your capacity larger. You grow and are willing to embrace more. Your memoir becomes richer."[24]

The "coherent life narrative" as a double-edged sword

In *Mindsight*, psychiatrist Daniel Siegel defines a life narrative as "the way we put our story into words to convey it to another person" and explains how mental health professionals elicit such narratives from their clients. "By simply asking certain kinds of autobiographical questions, we can discover how people have made sense of their past—how their minds have shaped their memories of the past to explain who they are in the present."[25] Although describing a diagnostic tool called the Adult Attachment Interview (AAI), he could just as easily be defining memoir, which Birkerts describes this way:

> Apart from whatever painful or disturbing events they recount, [memoirs'] deeper ulterior purpose is to discover the nonsequential

> connections that allow those experiences to make larger sense; they are about circumstance becoming meaningful when seen from a certain remove. They all, to greater or lesser degree, use the vantage point of the present to gain access to what might be called the hidden narrative of the past. Each is in its own way an account of detection, a realized effort to assemble the puzzle of what happened in the light of subsequent realization.[26]

"Detection" strikes me as a deliberate choice of word. Beyond simple "discovery," detection implies that meaning-making is an imaginative, deductive act, built from clues more than from direct inquisition or evidence. Interestingly, though, psychiatrists like Siegel seem more accepting of the primacy of emotional truth over documentable fact than are memoir writers and critics.

> The AAI evaluation accepts that memory is fallible. . . . Even at our most honest moments we say things we think others expect to hear, and we say them in ways that make us appear as we want to appear. For these reasons the analysis does not presume the accuracy of the facts as stated. Instead it focuses on the coherence of the story.[27]

This tension between "the accuracy of the facts" and "the coherence of the story" generates myriad challenges for the memoirists, who by virtue of choosing nonfiction—a genre defined by what it's *not*—hold themselves to a standard that may at times be unrealistically rigorous. Certainly there are cases of fiction masquerading as memoir for purely commercial reasons—James Frey's *A Million Little Pieces* comes to mind as an especially obvious and egregious example. But most memoirists would agree with Birkerts that "memoir, unlike reportage, serves the spirit of the past, not the letter. . . . The writer must represent as faithfully as possible what memory has shaped inside—memory and feeling."[28] Or, as author Nicole Walker so cogently observed, "What is creative nonfiction writing but the shaping and reshaping of self against fact?"[29]

Regardless of the ethics of labeling something "memoir" versus "fiction," Siegel believes that having a coherent life narrative—making sense of what has happened to you and of how you have come to be who you are—is a necessity for and marker of emotional and mental health, arguing that "making sense is a source of strength and resilience. In my twenty-five years as a therapist, I've also come to believe that making sense is essential to our well-being and happiness."

But Kabat-Zinn would maintain that this kind of narrative is a double-edged sword. At best, it is an illusion.

> [Both modern biology . . . and Buddhism would say that] . . . you will not find a permanent, independent, enduring self, whether you look for it in "your" body . . . in "your" emotions, "your" beliefs, "your" thoughts, "your" relationships, or anyplace else. And the reason you will not be able to locate anywhere a permanent, isolated, self-existing self that is "you" is that it is a mirage, a holographic emergence, a phantom, a product of the habit-bound, emotionally turbulent, thinking mind. It is being constructed and deconstructed continually, moment by moment. It is continually subject to change, and therefore not permanent or enduring or real.[30]

And at worst, Kabat-Zinn argues, having a coherent and *repeated* life narrative can imprison us, making us hapless victims of mental and emotional habits. "Without awareness of the anger or of the self-absorption, or ennui, or any other mind state that can take us over when it arises," he writes, "we reinforce those synaptic networks within the nervous system that underlie our conditional behaviors."[31]

In some respects, Patricia Hampl agrees with Kabat-Zinn, but questions whether we have any other choice:

> Each of us must possess a created version of the past. Created, that is, real in the sense of tangible, made of the stuff of a life lived in place and in history. And the downside of any created thing as well: We must live with a version that attaches us to our limitations, to the inevitable subjectivity of our point of view.[32]

Meredith Maran's memoir, *My Lie: A True Story of False Memory*, starkly illustrates how this quest to make sense of our feelings can lock us into a narrative that reinforces painful feelings, justifies dysfunctional ones, and has destructive consequences. She describes how she came to "remember" having been molested by her father, only to realize eight years after severing her own and her children's contact with him, that this was a false memory. Years later, she talks with her brother about her false (but, at the time, sincere) allegations.

> "The human species has a hard time achieving balance," he said. "It's hard to achieve balance when you're unhappy. You were unhappy. You were looking for balance. You thought you'd find it by putting the pieces into place, by remembering that you'd been abused. I can't blame you for looking for happiness."[33]

So is the quest to make meaning from memory and immediate experience an inherently dangerous pursuit? Or even an inherently worthy one? Without addressing that question, Hampl simply notes that it's a process that has consequences on a social as well as a personal level. She quotes novelist Milan Kundera, who writes, "The struggle of man against power is the struggle of memory against forgetting," then reinforces his point, flatly asserting that "what is remembered is what becomes reality." Hampl, through quoting Kundera, is talking about the tendency of those in power to create their own versions of the past, one that suits their current ideology, objectives, and actions. And she reminds us that "the function of memory, while experienced as intensely personal, is surprisingly political.[34]"

Living with paradox

If you were simply flipping through this essay, depending on which page you landed on, you might variously conclude that the "self" is an illusion or that the "sense of self" is what makes us human; that without the language center in the left brain we would have no such sense, or that self-awareness depends on the symbolic and pattern recognition capabilities of the right brain; that one's past feels fixed and ancient, or that memories inform and enhance our present awareness.

Writers in the realms of mindfulness, neuroscience, and memoir make all of these assertions, and all of them truthfully, if not comfortably, coexist. They converge in the recognition of a paradox beautifully articulated at the start of Paul Broks's *Into the Silent Land*:

> Reality is under constant review. . . . We still live by intuitions and illusions, especially when our thoughts turn inwards. The bright, intangible qualities of subjective experience have yet to be reconciled with the dark substance of the brain, but that space behind the face is still lit by the mind's eye. Irresistibly, we still see the vision of minds in the light of other people's eyes. Cosmologies come and go, but if this illusion begins to fade then so does the observer.[35]

Science, spirituality, and literature—we need all three to get through the day, the life, the universe.

Notes

Chapter 2

1 Droit-Volet, Sylvie. 2013. *Science Direct.* Vers. Volume 51, Issue 2. January. Accessed April 28, 2016.

2 Falk, Dan. 2013. *Do Humans Have a Biological Stopwatch.* January. Accessed October 22, 2015.

3 Eagleman, David. "BRAIN TIME." *BRAIN TIME | Edge.org.* Accessed April 28, 2016. https://www.edge.org/conversation/brain-time.

4 Eagleman, "BRAIN TIME," *BRAIN TIME | Edge.org.*

5 Falk, *Do Humans Have a Biological Stopwatch,* January.

6 Eagleman, "BRAIN TIME," *BRAIN TIME | Edge.org.*

7 Abbey, Edward. 1968. *Desert Solitaire.* New York: McGraw-Hill, pg. 202.

8 Abbey, *Desert Solitaire,* pg. 204–205.

9 Bowden, Charles. *Harper's Magazine.* 1998. "Torch Song." August. Accessed May 18, 2016.

10 Bowden, Charles. *Harper's Magazine.* 1998. "Torch Song." August. Accessed May 18, 2016.

11 Nelson, Jessica Hendry. 2014. *If Only You People Could Follow Directions.* Berkeley: Counterpoint Press, pg 54–55.

12 Nelson, *If Only You People Could Follow Directions,* pg 58–59.

13 Reece, Erik. 2006. *Lost Mountain.* New York City: Riverhead Books, pg 108–109.

14 Alison, Jane. 2009. *The Sisters Antipodes.* Boston: Mariner Books, pg 128.

15 Castro, Joy. 2012. *Island of Bones.* Lincoln: University of Nebraska Press, pg 28–129.

16 Finn, Charles. 2012. *Wild Delicate Seconds.* Corvallis: Oregon State University Press, 16–17.

17 Wilkins, Joe. 2012. *The Mountain and the Fathers.* Berkeley: Counterpoint Press, pg 112.

18 O'Brien, Tim. 1990. *The Things They Carried.* Boston: Mariner Books, pg 78.

19 O'Brien, *The Things They Carried*, pg 82.

20 O'Brien, *The Things They Carried*, pg 84.

Chapter 3

1 "'Do Not' psyches you out," Mitch grumbles. "'Do . . .' Alright! I get to disturb this guy. 'Not . . .' Shit! I need to read faster."

2 When I share this example with students, they tend to blush. "What?" I ask. "What connection have you made?" One of them eventually explains it like this: "Well, it sounds like a lactating woman who's asking her partner to milk her. Sugar is a term of endearment." Any more brain busters, Wittgenstein?

3 Since then, I've decided Shelley Jackson's "Ineradicable Stain" is the most experimental literary work-in-progress. Published one word at a time as tattoos on willing participants, the short story is only 25.4 percent published. The tattooed choose the temporary context ("It" on toe, "Dead" on scalp, "Blushes" on buttcheek) until all words are published and the permanent context, the story itself, is complete.

4 You be the judge: "The Gracehoper was always jigging ajog, hoppy on akkant of his joyicity."

Chapter 5

1 What Is an Image? W. J. T. Mitchell, *New Literary History*, Vol. 15, No. 3, Image/Imago/Imagination (Spring 1984), 503–37.

2 https://www.scientificamerican.com/article/when-we-read-we-recognize-words-as-pictures-and-hear-them-spoken-aloud/

3 https://www.princeton.edu/news/2011/08/31/word-association-princeton-study-matches-brain-scans-complex-thought

4 https://www.theguardian.com/science/2016/apr/27/brain-atlas-showing-how-words-are-organised-neuroscience

5 Ibid.

6 https://www.sciencedaily.com/terms/quantum_entanglement.htm

Chapter 8

1 Wikipedia has a thorough and reliable discussion of the lens.

2 *Salt Lake Tribune*, April 8, 1968.

3 "Good Eye."

4 Ibid.

5 Ibid.

6 "At Pompeii."

7 "Planetarium."

Chapter 9

1 Wolff, Tobias. Old School. New York: Knopf, 2003.

2 Wolff, Tobias. "Bullet in the Brain." *The Night in Question*. New York: Knopf, 1996.

3 Stern, Jerome. *Making Shapely Fiction*. New York: W. W. Norton, 1991.

4 Orwell, George. "Politics and the English Language." Accessed October 16, 2018. http://www.orwell.ru/library/essays/politics/english/e_polit

5 Bausch, Richard. *Facebook*, January 27, 2012. Accessed October 16, 2018. https://www.facebook.com/richard.bausch/posts/10151206746665075.

6 Kaye, Kenneth. "Controlling the Distance Between Reader & Character: The Primary Process." *The Writer's Chronicle*, February 2008.

7 Doidge, Norman. *The Brain That Changes Itself: Stories of Personal Triumph from the Frontiers of Brain Science*. New York: Penguin, 2007.

8 Doidge, Norman. *The Brain That Changes Itself: Stories of Personal Triumph from the Frontiers of Brain Science.* New York: Penguin, 2007.

9 Paul, Annie Murphy. "Your Brain on Fiction." *New York Times* (New York, NY), March 17, 2012. https://www.nytimes.com/2012/03/18/opinion/sunday/the-neuroscience-of-your-brain-on-fiction.html.

10 Plimpton, George. "Ernest Hemingway, The Art of Fiction No. 21." *The Paris Review*, Issue 18 (Spring 1958). Accessed October 16, 2018. https://www.theparisreview.org/interviews/4825/ernest-hemingway-the-art-of-fiction-no-21-ernest-hemingway

11 O'Connor, Flannery. "The Nature and Aim of Fiction." *Mystery and Manners.* New York: Farrar, Straus & Giroux, 1957. New York: Penguin, 2007.

12 Risen, Jane L. and Thomas Gilovich. "Why People Are Reluctant to Tempt Fate." *Journal of Personality and Social Psychology* 95, No. 2 (2008), 293–307. Accessed October 16, 2018. https://www.researchgate.net/publication/23136157_Why_People_Are_Reluctant_to_Tempt_Fate.

13 Conner, Roby. "On Smoking." Unpublished memoir. In the author's possession.

14 Conner, Roby. "On Smoking." Unpublished memoir. In the author's possession.

15 Conner, Roby. "On Smoking." Unpublished memoir. In the author's possession.

16 Bodwell, Joshua. "Stories We Love: Two Stories and a Life." *Fiction Writers Review*, May 24, 2012. https://fictionwritersreview.com/shoptalk/stories-we-love-two-stories-and-a-life/.

Chapter 11

1 I am indebted to the on-going study with my research partner, Hoke (Yao) Glover about metaphor, This essay represents the refinement of a portion our collective research, which led to pedagogical methods and approaches to reading and writing.

2 Letter to A. S. Lazarev, November 1, 1889. I. Ya. Gurlyand, in "Reminiscences of A. P. Chekhov," Teatr i Iskusstvo, July 11, 1904, states that Chekhov had told him the following in conversation at Yalta in the summer of 1889: "If in the first act you have hung a pistol on the wall, then in the following one it should be fired. Otherwise don't put it there."

Chapter 12

1 The people.

2 Language.

3 Beeplant; Cleome Serrulata.

4 Ponderos Pine; Pinus Ponderosa.

5 Sumac; Rhus Trilobata.

6 Sumac berries.

7 Plants.

8 Land of Diné.

9 Snowbowl, a business dedicated to inciting the vitriol that is often expressed by those who support Snowbowl. This vitriol is often in reply to the various tribes that repeatedly and unwaveringly communicate that the placement of reclaimed snow on Dook'o'oosłiid is an act of desecration allowed at the behest of racial discriminatory violence that has historically occluded tribes from not only taking any part in the region's decision-making processes but has led to mass thefts of land.

10 Having to do or taking place among the Diné.

11 Another colonial namesake that dates back to the Spanish conquerors, patterning the same interactions that stem from languages known to lead into radical departures from Dinétah.

12 Lake Powell is a dam constructed by the state of Arizona to generate hydroelectric power. This hydroelectric power has been attained with use of coal extracted from Dził Yizhiin (Black Mesa). Both the obstruction of the river and the extraction of coal have negatively impacted Dinétah in the form of persistent droughts, relocation of Diné who have historic ties to lands, and a degradation in air quality. Arizona's public and private utilities—now some of the largest energy companies in the world—dispense this electricity to Tucson, Phoenix, Las Vegas, and California. The water that trickles from the dam is then channeled through Havasu Falls, to Phoenix and Tucson, for general water use without any placements of water restriction use.

13 Much like a cornstalk, my ancestors' emergence began when they climbed from the previous world into this world, an era that's called the Glittering World, aka Ni' Hodisǫs.

14 The paradigm that we currently live in.

15 The walk, known at the time as Nihígáál bee Iiná (Our Journey for Existence), took place in 2015, in which walkers walked ~2,000 miles. This specific walk involved Diné youth engaging with various communities across Dinétah to address the adverse effects of resource extraction (e.g., fracking, coal mining, uranium, and water depletion) while revitalizing cultural practices. While Nihígáál bee Iiná is an expression of activism, it also has its roots in a traditional practice in which Diné people, both of the past and the present, make pilgrimages to each of our sacred mountains to give offering.

16 Womenfolk.

17 Maternal grandmother.

18 Deer Spring people.

Chapter 13

1 This paragraph was inspired by Nabokov's descriptions of butterflies in *Speak Memory, An Autobiography Revisited* (New York: G.P. Putnam and Sons, 1966).

2 Vladimir Nabokov, *Speak Memory: An Autobiography Revisited* (New York: G.P. Putnam and Sons, 1966), 124.

3 Ibid., 125.

4 Ibid.

5 Kurt Johnson and Steve Coates, *Nabokov's Blues: The Scientific Odyssey of a Literary Genius* (Cambridge: Zoland Books, 1999), 319.

6 Vladimir Nabokov, *Strong Opinions* (New York City: McGraw Hill International, 1973), 10.

7 Johnson and Coates, *Nabokov's Blues*, 309.

8 Nabokov, *Speak Memory*, 119.

9 Johnson and Coates, *Nabokov's Blues*, 306.

10 *New York Times Book Review*, December 28, 1952.

11 Nabokov, *Speak Memory*, 139.

Chapter 16

1 Simon McCrea, "Intuition, Insight, and the Right Hemisphere: Emergence of Higher Sociocognitive Functions," *Psychology, Research, and Behavior Management* 3 (2010): 1–39. www.ncbi.nlm.nih.gov

2 Carl Klaus and Ned Stuckey-French, *Essayists on the Essay: Montaigne to Our Time*, 1. (University of Iowa Press, 2012).

Chapter 17

1 Interview on RadioLab.

2 Sven Birkerts, *Then, Again: The Art of Time in Memoir* (Minneapolis: Graywolf Press, 2008), 8.

3 Jon Kabat-Zinn, *Coming to Our Senses* (New York: Hyperion Press, 2005), 43.

4 Patricia Hampl, *I Could Tell You Stories: Sojourns in the Land of Memory* (New York: W.W. Norton and Company, 1999), 32.

5 Kabat-Zinn, *Coming to Our Senses*, 168.

6 Hampl, *I Could Tell You Stories*, 33.

7 Kabat-Zinn, *Coming to Our Senses*, 9.

8 Joan Didion, *The White Album* (New York: Farrar, Strauss, and Giroux, 1979), 205.

9 Vivian Gornick, *Fierce Attachments* (New York: Farrar, Strauss, and Giroux, 1987), 14.

10 Ibid., 48.

11 Kabat-Zinn, *Coming to Our Senses*, 168.

12 Jill Bolte Taylor, "Jill Bolte Taylor's Stroke of Insight," *TED Talk*, February 2008, posted March 2008. http://www.ted.com/index.php/talks/jill_bolte_taylor_s_powerful_stroke_of_insight.html

13 Paul Broks, *Into the Silent Land: Travels in Neuropsychology* (New York: Atlantic Monthly Press, 2004), 17.

14 Kabat-Zinn, *Coming to Our Senses*, 326.

15 V. S. Ramachandran, *The Tell-Tale Brain: A Neuroscientist's Quest for What Makes us Human* (New York: W.W. Norton and Company, 2011), Kindle location 4626.

16 Mary Karr, *The Art of Memoir* (New York: Harper Collins, 2015), 23.

17 Birkerts, *Then, Again*, 43.

18 Abigail Thomas, *A Three Dog Life* (New York: Harcourt, Inc., 2006), 3.

19 Anne Lamott, *Bird by Bird* (New York: Anchor Books, 1994), 123.

20 David Sedaris, "Repeat After Me." From Lex Williford and Michael Martone (ed.), *Touchstone Anthology of Contemporary Creative Nonfiction* (New York: Touchstone, 2007), 446.

21 Kabat-Zinn, *Coming to Our Senses*, 162–63.

22 Virginia Woolf, *Moments of Being* (Orlando: Harcourt, Brace and Company, 1945), 98.

23 Kabat-Zinn, *Coming to Our Senses*, 163.

24 Natalie Goldberg, *Old Friend from Far Way: The Practice of Writing Memoir* (New York: Free Press, 2007), Kindle location 618–24.

25 Daniel Siegel, *Mindsight* (New York: Bantam Books, 2012), 171.

26 Birkerts, *Then, Again*, 8.

27 Siegel, *Mindsight*, 174.

28 Birkerts, *Then, Again*, 142.

29 Nicole Walker, "The Braided Essay as Social Justice Action," *Creative Nonfiction*, Issue #64, https://www.creativenonfiction.org/online-reading/braided-essay-social-justice-action

30 Kabat-Zinn, *Coming to Our Senses*, 326.

31 Ibid., 71.

32 Hampl, *I Could Tell You Stories*, 33.

33 Meredith Maran, *My Lie: A True Story of a False Memory* (San Francisco: Jossey-Bass, 2010), 236.

34 Hampl, *I Could Tell You Stories*, 33.

35 Broks, *Into the Silent Land*, 3.

Contributors

Nancer Ballard is a fiction and nonfiction writer, poet, lawyer, and resident scholar at Brandeis University's Women's Studies Research Center. She studied the cognitive science of memory with the late John Lisman at Brandeis University and is working on a memoir about growing up with a mother who could not produce visual imagery—a condition known as "Aphantasia." She is the author of "Dead Reckoning," a collection of poems, and is coauthor of a supplemental children's textbook series that teaches mathematics through multicultural-based storytelling. Her recent creative nonfiction has appeared in *Here Comes Everyone*, *Memoryhouse*, *Thema Literary Journal*, *South Street Literary Journal*, and numerous anthologies including *The Far Edges of the Fourth Genre, An Anthology of Explorations in Creative Nonfiction*, and *Affective Disorder and the Writing Life: The Melancholic Muse.*

Lyncia Begay is a newly emerging Diné writer and artist based out of so-called Arizona who creates nonfiction pieces, acutely focused on the impacts of colonization. Her work often features elements of prose, poetry, essay, and story.

In this essay, her work offers an integration of indigenous philosophy to offer the similitude that knowledge systems often share. Throughout this essay, there is a constant stream of themes that generously offers a series of humanizing moments in what is so often viewed as inhuman. Her ideas cause readers to challenge their perceptions of water and how they relate and honor the continuity of the water ways that are exceedingly disappearing in response to the sixth extinction.

Lyncia's interests consist of creating community projects geared toward cultural revitalization and indigenous education systems. Lyncia has also graduated with a masters in Creative Writing from Northern Arizona University.

Frank Bures is the author of *The Geography of Madness: Penis Thieves, Voodoo Death and the Search for the Meaning of the World's Strangest Syndromes*, in which he argues that culture is a narrative ecosystem that we share with others. His work has been included in the *Best American Travel Writing* and selected as "notable" in the *Best American Essays* and *Best American Sports Writing*. More at frankbures.com.

J.T. Bushnell teaches writing and literature at Oregon State University. His fiction and creative nonfiction have appeared in *Passages North, The Mississippi Review*, *Iron Horse Literary Review*, *North Dakota Quarterly*, *Flyway*, *Monkeybicycle*, *New Madrid*, *Meridian*, *The Greensboro Review*, *The South Carolina Review*, *Brevity*, and other journals. He also contributes essays about writing to *Poets & Writers*, *The Writer*, and *Fiction Writers Review*, where he has served as a contributing editor. He lives in Eugene, Oregon.

Katharine Coles's seventh collection of poems *Wayward* was released in 2019; in 2018, she published a memoir *Look Both Ways*. In the United States, she has received awards from the National Endowment for the Arts, and National Science Foundation, the National Endowment for the Humanities, and the Guggenheim

Foundation. She is also the founding co-Director of the Utah Symposium in Science and Literature, the Poet-in-Residence for the Poet's House Field Work Program, and on the editorial board of the Journal of Dark Skies Studies. She is a distinguished professor of English at the University of Utah.

David Lazar's books include *I'll Be Your Mirror: Essays and Aphorisms, Who's Afraid of Helen of Troy, After Montaigne, Occasional Desire: Essays, The Body of Brooklyn, Truth in Nonfiction, Essaying the Essay, Powder Town, Michael Powell: Interviews,* and *Conversations with M.F.K. Fisher*. His forthcoming books are Celeste Holm Syndrome, a book of essays, and *Don't Look Now*, an anthology. Nine of his essays have been "Notable Essays of the Year" according to *Best American Essays*. He established the PhD and MA programs in nonfiction writing at Ohio University, and currently he is Professor of creative writing at Columbia College, Chicago, where he directed the creation of the MFA program in nonfiction. Lazar is founding editor of the literary magazine *Hotel Amerika*, now in its eighteenth year, and series coeditor, with Patrick Madden, of *21st Century Essays*, at Ohio State University Press. Lazar was a Guggenheim fellow in nonfiction for 2015–16.

Lawrence Lenhart studied writing at the University of Pittsburgh and holds an MFA from the University of Arizona. His essay collections are *The Well-Stocked and Gilded Cage* (Outpost19) and *Of No Ground: Small Island/Big Ocean Contingencies* (West Virginia University Press). His prose appears in *Creative Nonfiction, Fourth Genre, Gulf Coast, Passages North, Prairie Schooner*, and elsewhere. He is the Associate Chair of the English Department at Northern Arizona University where he also teaches fiction, nonfiction, and climate science narratives. Lenhart is a founding editor of *Carbon Copy* and the reviews editor of *Diagram*.

Dave Madden is the author of *The Authentic Animal: Inside the Odd and Obsessive World of Taxidermy* and the story collection, *If You Need Me I'll Be Over There.* He received fellowships from the MacDowell Colony and the Bread Loaf Writers' Conference, and he teaches in the MFA program at the University of San Francisco.

Leila Philip is the author of four books, including *The Road through Miyama*, which received the 1990/PEN Martha Albrand Special Citation for Nonfiction, and *A Family Place: A Hudson Valley Farm, Three Centuries, Five Wars, One Family*, which received awards for history and documentation of American life. Philip has received numerous honors for her writing, including from the John Simon Guggenheim Foundation, the National Endowment for the Arts, and the National Endowment for the Humanities. In addition to publishing essays and poems, Philip is a contributing columnist at *The Boston Globe*. Her most recent book is *Water Rising* (2015) with Garth Evans, an art collaboration which has led to musical performances and the creation of a video installation, "Environment, Memory & Things," which has been shown internationally. *Water Rising* has an environmental mission; 100 percent of proceeds from sales of the book are being donated to support environmental stewardship. She teaches at the College of the Holy Cross where she is a professor of literature and creative writing in the English department.

V. Efua Prince's current research titled *Waterbearers* tells a compelling story about the convergence between unpaid and underpaid labor and industry, of housework and industrial development, and of generations of African American women whose domestic service is transformed into a labor of love. Prince's first monograph *Burnin' Down the House: Home in African American Literature* was published in 2005. Her second monograph *Daughter's Exchange* is a hybrid text exploring the African American woman's encounter with the intellectual marketplace. Prince has been a fellow at Harvard University's W. E. B. Du Bois Institute, a visiting scholar at the University of Virginia's Carter G. Woodson Institute, and the Avalon Professor of Humanities at Hampton University. Currently, she is an associate professor of African American Studies at Wayne State University.

Ira Sukrungruang is the author of three nonfiction books *Buddha's Dog & Other Mediations*, *Southside Buddhist*, and *Talk Thai: The Adventures of Buddhist Boy*; the short story collection *The Melting Season*; and the poetry collection *In Thailand It Is Night*. He is president of *Sweet: A Literary Confection* (sweetlit.com) and is the Richard L. Thomas Professor of Creative Writing at Kenyon College.

Marco Wilkinson's work has appeared in *Kenyon Review*, *Territory*, *Seneca Review*, *Diagram*, *Bennington Review*, and elsewhere. He teaches writing, environmental literature, and sustainable agriculture at Oberlin College and Lorain County Community College in Ohio. A companion piece "A Gardener's Education (Plant Body)" to his essay here was published in the online journal *Territory*. His memoir, *Madder*, is forthcoming from Coffee House Press in 2021.

Julie Wittes Schlack is the author of the linked essay collection, *This All-at-Onceness: A Memoir of Hope and Satellites*. She has an MFA in creative writing from Lesley University. Her work has appeared in numerous literary journals such as *Shenandoah*, *The Writer's Chronicle*, *Ninth Letter*, and *The Tampa Review*. She is a former book reviewer for *The Boston Globe*, currently reviews books for *The ARTery*, and is a regular contributor to *Cognoscenti*, National Public Radio station WBUR's journal of ideas and opinions. Julie lives in Cambridge, Massachusetts, with her husband. You can read more of her work at www.juliewittesschlack.com.

Amy Wright is the author of two poetry books, one poetry collaboration, and six chapbooks, including the essay chapbook *Think I'll Go Eat a Worm*. Most recently, her nonfiction won first place in the Writers at Work essay contest. Additional essays are published in *Brevity, Kenyon Review, Ninth Letter, Waveform: Anthology of Women Essayists*, and elsewhere.

Editors

Sean Prentiss is the award-winning author of *Finding Abbey: The Search for Edward Abbey and His Hidden Desert Grave*, winner of the 2015 National Outdoor Book Award. Prentiss is also the author of *Environmental and Nature Writing: A Writer's Guide and Anthology* and the author of *Crosscut: Poems*. He and his family live on a small lake in northern Vermont, and he serves as an associate professor at Norwich University and is a core faculty member in the MFA Writing and Publishing program at Vermont College of Fine Arts.

Nicole Walker is most recently the author of *Sustainability: A Love Story* and *The After-Normal*. Her previous books include *Where the Tiny Things Are, Egg, Micrograms, Quench Your Thirst with Salt*, and *This Noisy Egg*. She also edited *Bending Genre: Essays on Creative Nonfiction* with Margot Singer. She's nonfiction editor at *Diagram* and professor at Northern Arizona University in Flagstaff.

Acknowledgments

The following pieces have been previously published.

Frank Bures: "The Secret Lives of Stories: Rewriting Our Personal Narratives" first appeared in *Poets & Writers Magazine*.

JT Bushnell: "The Heart and the Eye: How Description Can Access Emotion" essay was originally published in *Poets & Writers*, May/June 2016.

Katherine Coles: "Lens: A Lyric Meditation" was originally published in *New Writing*, vol. 15, November 2017.

Dave Madden: "The Brain Is a Master Class" was first published in a different form in *Creative Nonfiction* #61, Fall 2016."

Sean Prentiss: "Bengal Tiger Moments: Perception of Time in the Brain and on the Page" was originally published in *The Writer's Chronicle*, vol. 50, number 2, December 2017.

Julie Wittes Schlack: "Mindfulness and Memoir" was originally published in *The Writer's Chronicle*, vol. 46, number 3, December 2013.

Bibliography

Chapter 7

Aristotle. *On Sense and the Sensible* [DeSensu et Sensibilibus], trans. J. I. Beare. Clarendon Press, 1931. https://ebooks.adelaide.edu.au/a/aristotle/sense/ (July 8, 2018).

Aziz-Zadeh, Lisa, Stephen M. Wilson, Giacomo Rizzolatti, and Marco Iacoboni. "Congruent Embodied Representations for Visually Presented Actions and Linguistic Phrases Describing Actions," *Current Biology* 16 (2006): 1818–23.

Brox, Jane. *Clearing Land; Legacies of the American Farm*. New York: North Point Press, 2004.

Casasanto, Daniel, Roel Willems, and Peter Hagoort. "Body-specific Representations of Action Verbs: Evidence from fMRI in Right-and Left-handers," *Proceedings of the Annual Meeting of the Cognitive Science Society* 31 (2009): 875–80.

Cook, Andrew. "Exploding the Five Senses, a Sideways Look at Human Sensing Abilities," June 30, 2000. http://www.hummingbird-one.co.uk/resources/archive/five.html (July 1, 2018).

Dekker, Tessa M., Denis Mareschal, Mark H. Johnson, and Martin I. Sereno. "Picturing Words? Sensorimotor Cortex Activation for Printed Words in Child and Adult Readers," *Brain Language* 139 (December 2014): 58–67.

Fisher, M. F. K. *As They Were*. New York: Vintage Books, 1983.

Grealy, Lucy. *Autobiography of a Face*. Boston: Houghton Mifflin, 1994.

Hoffman, Donald D., qtd. by Amanda Gefter, "The Case Against Reality," *The Atlantic*, April 25, 2016. http://www.theatlantic.com/science/archive/2016/04/the-illusion-of reality/479559

Iacoboni, Marco. *Mirroring People*. New York: Picador, 2009.

Krakauer, John. *Into Thin Air*. New York: Villard, 1997.

Kuzmičová, Anežka. *Mental Imagery in the Experience of Literary Narrative: Views from Embodied Cognition* (diss.). Stockholm: Stockholm University, 2013.

Meloy, Ellen. "Tilano's Jeans," in *The Anthropology of Turquoise*. New York: Vintage Books, 2002, 163–70.
Orwell, George, "Such, Such Were the Joys…," in *The Art of the Personal Essay*, ed. Phillip Lopate. New York: Anchor Books, 1995, 269–302.
Proffitt, Dennis R. qtd. by Christof Koch. "Looks Can Deceive: Why Perception and Reality Don't Always Match Up," *Scientific American Mind*, July 1, 2010. http://www.scientificamerican.com/article/looks-can deceive/
Shenk, Joshua Wolf. *Lincoln's Melancholy; How Depression Challenged a President and Fueled his Greatness*. Boston: Houghton Mifflin Company, 2005.
Skloot, Rebecca. *The Immortal Life of Henrietta Lacks*. New York: Broadway Paperbacks, 2011.
Smith, Barry, qtd. in Joe Humphreys. "Aristotle Got It Wrong: We Have a Lot More than Five Senses," *The Irish Times*, May 16, 2017. https://www.irishtimes.com/culture /aristotle-got-it-wrong-we-have-a-lot-more-than-five-senses-1.3079639 (July 5, 2018).
Smith, J. David, Margaret Wilson, and Daniel Reisberg. "The Role of Subvocalization in Auditory Imagery," *Neuropsychologia* 33, no. 11 (1995): 1433–54.
Speer, Nicole K., Jeremy R. Reynolds, Khena M. Swallow, and Jeffrey M. Zacks. "Reading Stories Activates Neural Representations of Visual and Motor Experiences," *Psychological Science* 8 (August 20, 2009): 989–99.
Stern, Daniel. *The Present Moment in Psychotherapy and Everyday Life*. New York: WW Norton & Company, 2004.
Wallace, Alfred Russell. *Tropical Nature and Other Essays*, qtd. in David Grambs and Ellen S. Levine. *The Describer's Dictionary*. New York: W.W. Norton & Company, 2015, 179–80.

Chapter 8

Coles, Katharine. "Good Eye," in *Fault*. Los Angeles: Red Hen Press, 2008, 12–15.
Coles, Katharine. "At Pompeii," in *Flight*. Los Angeles: Red Hen Press, 2016, 21.
Fabrizio, Doug. "Host and Interviewer," *Radio West*. http://radiowest.kuer.org/post/conversation-katharine-coles (April 4, 2016).
Rich, Adrienne. "Planetarium," in *Collected Poems*. New York: W.W. Norton, 2016, 301–2.
Johnson, Lynn. Tribune staff photographer. Salt Lake Tribune. April 8, 1968.
Wikipedia. https://en.wikipedia.org/wiki/Lens_(anatomy) (Accessed June–July 2017).

Chapter 10

Bartlett, Sir Frederic. *Remembering: A Study in Experimental and Social Psychology*. Cambridge: Cambridge University Press, 1995.

Bergson, Henri. *Matter and Memory*. Cambridge: MIT Press, 1988 (originally published 1911). Casey, E. *Remembering*. A phenomenological study. Bloomington: Indiana University Press, 2000.

Chabris, Charles, and Simons, Daniel. "What People Believe about How Memory Works: A Representative Survey of the U.S. Population," *PlosOne* 6, Issue 8 (2011): 1–7.

Draaisma, Douwe. *Metaphors of Memory: A History of Ideas about the Mind*. Cambridge: Cambridge University Press, 2000.

Ebbinghaus, Hermann. *Memory: A Contribution to Experimental Psychology*. New York: Columbia University, 1913.

Fernyhough, Charles. *Pieces of Light: How the New Science of Memory Illuminates the Stories We Tell About Our Pasts*. New York: Harper Perennial, 2012.

Fuchs, T. *Das Gedächtnis des Leibes* [The memory of the body]. Phänomenologische Forschungen 5, no. 1 (2000): 71–89.

Fuchs, T. *The Body and the Life-World*. New philosophical-psychiatric essays. Amsterdam, Netherlands: John Benjamins Publishing Company, 2012.

Fuchs, T. "Body Memory: An Integration," in *Leibgedächtnis und Lebensgeschichte* [Body memory and life history], ed. F. A. Friedrich, T. Fuchs, J. Koll, B. Krondorfer, and G. M. Martin. Der Text im Körper, University of Warwick, 2008.

Fuchs, T., S. Koch, and M. Summa *Leibgedächtnis, Inkarnation und Bibliodrama* [*The Text in the Body. Body-Memory, Incarnation and Bibliodrama*]. Hamburg: EB-Verlag, 2012, 10–40.

Fuchs, T., S. Koch, and M. Summa "Body Memory and Kinesthetic Body Feedback: the Impact of Light vs. Strong Movement Qualities on Affect and Cognition," *Memory Studies* 5 (June 2014): 1–29.

Hartley, L. P. *The Go-Between*. New York: New York Review of Books Classic Reprint, 2002.

Koch, S., T. Fuchs, M. Summa, and C. Müller (eds.). *Body Memory, Metaphor and Movement*. Amsterdam and Philadelphia: John Benjamins, 2012.

Laurin, Kristin. "Inaugurating Rationalization: Three Field Studies Find Increased Rationalization When Anticipated Realities Become Current," *Association for Psychological Science* (February 2018): 483–95.

Loftus, Elizabeth. *The Myth of Repressed Memory*. New York: St. Martin's Press, 1992.

Sacks, Oliver. *The River of Consciousness*. New York: Alfred Knopf, 2017.
Schwartz, Delmore. "In Dreams Begin Responsibilities," in *In Dreams Begin Responsibilities and Other Stories*, ed. James Atlas. New York: New Directions, 1978, 1–9.
Tulving, Endel. *Elements of Episodic Memory*. Oxford: Oxford University Press, 1985.
Tulving, Endel and Fergus Craik. *The Oxford Handbook of Memory*. Oxford: Oxford University Press. 2000.
Whitehead, Anne and Micheal Rossington. *Theories of Memory: A Reader*. Baltimore: Johns Hopkins University Press, 2007.

Chapter 11

Brown, Richard H. *A Poetic for Sociology: Toward a Logic of Discovery for the Human Sciences*. Chicago: University of Chicago Press, 1989. Print.
Holloway, John. "Narrative Structure and Text Structure: Isherwood's 'A Meeting by the River,' and Muriel Spark's 'The Prime of Miss Jean Brodie,'" *Critical Inquiry* 1, no. 3 (March 1975): 581–604. http://www.jstor.org/stable/1342832 (Accessed May 4, 2017).
Lorde, Audrey. "The Master's Tools Will Never Dismantle the Master's House," in *Sister Outsider: Essays and Speeches*. New York: Quality Paperback Club, 1993, 110–13.
Morrison, Toni. *The Bluest Eye*. New York: Vintage, 2007. 1970.
Upton, Brian. *The Aesthetic of Play*. Cambridge: MIT Press, 2015. http://www.jstor.org/ stable/j.ctt17kk75j.18 (Accessed May 8, 2017).

Index

CPSIA information can be obtained
at www.ICGtesting.com
Printed in the USA
LVHW080351090222
710649LV00002B/42

9 781350 083882